Whitewares and Materials

Ceramic Engineering & Science Proceedings Volume 25, Issue 2, 2004

Series Editor: Greg Geiger
Production Manager: John Wilson
Director, Technical Publications: Mark Mecklenborg

Editorial and Circulation Offices
PO Box 6136
Westerville, Ohio 43086-6136

Contact Information
Editorial: (614) 794-5858
Customer Service: (614) 794-5890
Fax: (614) 794-5892
E-Mail: info@ceramics.org
Website: www.ceramics.org/cesp

Ceramic Engineering & Science Proceedings (CESP) (ISSN 0196-6219) is published five times a year by The American Ceramic Society, PO Box 6136, Westerville, Ohio 43086-6136; www.ceramics.org. Periodicals postage paid at Westerville, Ohio, and additional mailing offices.

Change of Address: Please send address changes to *Ceramic Engineering and Science Proceedings*, PO Box 6136, Westerville, Ohio 43086-6136, or by e-mail to info@ceramics.org.
Subscription rates: One year $220 (ACerS member $176) in North America. Add $40 for subscriptions outside North America. In Canada, add GST (registration number R123994618).
Single Issues: Single issues may be purchased online at www.ceramics.org or by calling Customer Service at (614) 794-5890.
Back Issues: When available, back issues may be purchased online at www.ceramics.org or by calling Customer Service at (614) 794-5890.

Indexing: An index of each issue appears at www.ceramics.org/ctindex.asp.
Contributors: Each issue contains a collection of technical papers in a general area of interest. These papers are of practical value for the ceramic industries and the general public. The issues are based on the proceedings of a conference. Both The American Ceramic Society and non-Society conferences provide these technical papers. Each issue is organized by an editor, who selects and edits material from the conference proceedings. The opinions expressed are entirely those of the presenters. There is no other review prior to publication. Author guidelines are available on request.
Postmaster: Please send address changes to *Ceramic Engineering and Science Proceedings*, P.O. Box 6136 Westerville, Ohio 43086-6136. Form 3579 requested.

Ceramic Engineering & Science Proceedings Volume 25, Issue 2, 2004

Whitewares and Materials

A collection of Papers Presented at the 105th Annual Meeting of The American Ceramic Society and the Whitewares and Materials Division Fall Meeting, held in conjunction with ACerS Canton-Alliance Section and the Ceramic Manufacturer's Association.

William M. Carty
Editor

April 27–30, 2003
Nashville, Tennessee
and
October 15–16, 2003
Mansfield, Ohio

Published by
The American Ceramic Society
735 Ceramic Place
Westerville, OH 43081
www.ceramics.org

ISSN 0196-6219

ISBN: 978-0-470-05147-4

Contents

Whitewares and Materials

105th Annual Meeting of The American Ceramic Society and the Whitewares and Materials Division Fall Meeting.

Preface

This issue of *Ceramic Engineering and Science Proceedings* contains papers and abstracts presented during sessions at the 105th Annual Meeting of The American Ceramic Society (ACerS), April 27–30, 2003, Nashville, Tennessee and the ACerS Whitewares and Materials Division Fall Meeting, held in conjunction with the Ceramic Manufacturers' Association (CerMA) and the ACerS Canton-Alliance Section, October 15–16, 2003, Mansfield, Ohio.

William M. Carty, Ph.D.
NYS CACT—Whiteware Research Center
New York State College of Ceramics at Alfred University

Observations of Matte Glaze Formation Independent of Firing Cycle

M. E. Katz, B. Quinnlan, W. M. Carty, and T. Gebhart
New York State College of Ceramics at Alfred University, Alfred, New York

Data pertaining to the compositional properties of matte glaze formation independent of cooling cycle properties are presented. The study examines several compositional variations to foster understanding of the formation of crystalline materials in glazes as a condition of glaze composition, as opposed to firing cycle and cooling.

Effect of Zinc Oxide Addition on Crystallization Behavior and Mechanical Properties of a Porcelain Body

S.-K. Kim, S.-M. Lee, E.-S. Choi, and H.-T. Kim
Pottery/Structural Ceramics Center, Korea Institute of Ceramic Engineering and Technology, Seoul, Korea

The effect of ZnO addition on crystallization behavior and mechanical properties of porcelain body has been investigated. As ZnO content increased, gahnite ($ZnAl_2O_4$) phase developed and cristobalite formation was promoted. However, alumina added as a raw material remained almost intact during sintering. The gahnite crystallized from 1130°C, after feldspar was significantly melted and ZnO dissolved into the glass. After sintering, abundant gahnite crystals with sizes of 50–400 nm were homogenously distributed inside glass phase. With increases in ZnO content and sintering temperature, water absorption decreased and flexural strength improved. The strength relies on ZnO content rather than water absorption. Wear resistance was enhanced with ZnO content.

Introduction

Zinc oxide has been known as a strong flux in pottery industries, and is used as one of the major components in commercial glazes for sanitaryware and tile.[1–3] It controls thermal expansion coefficient, reduces glaze viscosity, and enhances surface gloss. If it is used at more than 10 wt% in glaze, crystalline phase develops and the glaze has a matte appearance.[4] When much higher amounts are employed and the appropriate cooling cycle is adopted, decorative large willemite (Zn_2SiO_4) crystals appear on the surface of glaze; this effect has been applied to many artistic works.[5,6]

In crystalline glazes and glass ceramics containing ZnO, willemite and gahnite formation have been widely investigated.[7–13] The crystallization depends on both the chemical composition and the heat treatment cycle.[7–9,11] Sometimes, coloring agents such as cobalt and copper oxide are introduced to impart color and control crystallization.[9] While the willemite crystal shape changes from rodlike to a fine, equiaxed shape, depending on the annealing temperature, gahnite always has a starlike shape.[7–9] In addition, gahnite crystallization has been studied to enhance the wear resistance of tile glaze, because gahnite has spinel structure and hence high hardness and scratch resistance (8.5 on the Mohs scale).[10] Concerning the crystallization temperature, the gahnite started to form at about 1000°C in the SiO_2-B_2O_3-Al_2O_3-ZnO-CaO system[10] and at 900°C in the SiO_2-B_2O_3-Al_2O_3-ZnO-Li_2O system.[11]

In spite of many investigations of ZnO effects on crystallization in glaze and glass ceramics, studies of its effects on the porcelain body are very rare. Compared to glaze and glass ceramics, where all starting materials are usually completely melted and then crystallized during heat treatment at lower temperatures, the porcelain body consists of glass and many crystalline phases with complex microstructure during heating and isothermal holding. The glass phase in a porcelain body is initially provided by raw materials with low melting temperature such as feldspar. Other raw materials dissolve into the glass and then new crystalline phases develop. Therefore, the crystallization behavior of the porcelain body with ZnO addition must be different from those of glaze and glass ceramics.

In this study, the effect of ZnO addition on the crystallization and mechanical properties of a porcelain body was investigated. With an increase in ZnO content, the phase development of porcelain body has been examined and the mechanism of crystallization has been proposed. Then, water absorption and flexural strength were measured with ZnO content and sintering temperature. Finally, the potential application of this material was discussed.

Experimental Procedures

The chemical compositions of raw materials and their proportioned ratio for the reference specimen are listed in Tables I and II, respectively. To investigate the effect of ZnO addition, 1, 3, and 5 wt% ZnO was added to the batch for the reference specimen. The proportioned powders were ball milled for 2 h in a porcelain jar with a purified water medium. The milled slurries were passed through a 325-mesh sieve and then dried in oven at 90°C for 24 h. The dried powders were compacted and pressed into 4 mm × 5 mm × 40 mm pellets under 50 MPa. The pressed specimens were sintered at temperatures between 1000 and 1300°C and then furnace cooled. The heating rate to isothermal holding temperature was 7°C/min, and the cooling rate to 1000°C was 10°C/min. To check the possibility of crystallization during cooling, a specimen was water quenched and analyzed.

The crystalline phases present in the specimens were identified via X-ray diffractometry (XRD) with CuK_{α} target. The water absorption of the sin-

Table I. Chemical compositions (wt%) and LOI of raw materials

	SiO_2	Al_2O_3	Fe_2O_3	CaO	MgO	K_2O	Na_2O	TiO_2	LOI
Clay	47.12	34.39	1.36	0.21	0.23	1.2	0.41	0.45	14.63
Kaolin	44.64	37.1	1.77	0.6	0.48	0.64	0.45	0.23	14.09
Feldspar	75.6	14.0	0.1	0.35	0.02	3.87	5.32		0.74
Pyrophyllite	72.11	19.45	0.1	0.11	0.05	0.1	0.38	0.2	7.50
Alumina	0.02	99.7	0.01				0.26		

tered specimen was measured according to ASTM C373.[14] The flexural strength of the specimen was measured through the three-point bend test, which was conducted with a crosshead speed of 0.5 mm/min and an outer span of 30 mm. For the measurement, the specimens were polished to 3 μm and chambered with a diamond disk. For a few specimens, thermal scanning calorimetric (DSC) analysis was performed from room temperature to 1300°C. To observe the microstructure through scanning electron microscopy (SEM), the sintered specimen was prepared by polishing to 3 μm and etching in 10% HF solution for 1 min. The specimen was also analyzed through transmission electron microscopy(TEM) and energy dispersive X-ray spectroscopy (EDX). To estimate the wear resistance of the specimen, weight loss was measured when 10 kg of 20-mesh SiC particles were poured into the sintered specimen with an inclination angle of 45°.[15] Smaller weight loss indicates higher wear resistance.

Table II. Proportioned ratio for the reference specimens (wt%)

Clay	20.0
Kaolin	20.0
Feldspar	14.4
Pyrophyllite	28.1
Alumina	17.5

Results and Discussion

Figure 1 shows the XRD patterns of the specimens sintered at 1300°C for 30 min with an increase in ZnO addition. The addition of ZnO resulted in the development of a new phase, which was identified as $ZnAl_2O_4$ (gahnite, JCPDS powder diffraction file card no. 05-0669). In addition, cristobalite crystallization seems to be promoted with ZnO addition. On the contrary, the diffraction intensities from alumina remained almost constant, implying that the alumina added as a raw material didn't react with other materials. To check whether the crystalline phases developed during cooling from the

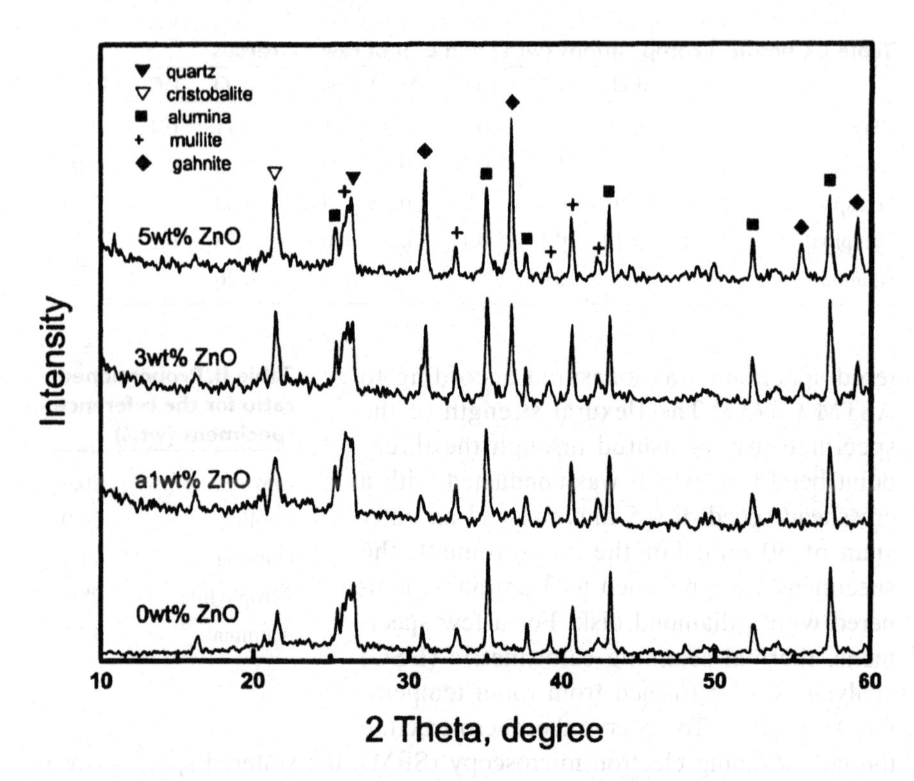

Figure 1. X-ray diffraction patterns of specimens with ZnO addition sintered at 1300°C for 30 min.

sintering temperature, the specimen with 5 wt% ZnO was water quenched from the sintering temperature of 1250°C after 1 h of isothermal holding, and then compared to specimens furnace-cooled after 1 min and 1 h of isothermal holding (Fig. 2). The diffraction pattern of the quenched specimen was the same with that of furnace-cooled specimen, while an increase in sintering time resulted in stronger diffraction intensities of gahnite and cristobalite. Therefore, the crystallization must occur during the isothermal holding at the soaking temperature rather than cooling period.

Figure 3 shows the diffraction patterns when the specimens with 3 wt% ZnO were sintered for 30 min as the sintering temperature is raised from 1000 to 1250°C. Up to 1000°C, ZnO remained and started to disappear with the melting of feldspar at increased sintering temperature. When feldspar was mostly melted at 1150°C, gahnite and cristobalite phases started to develop. The DSC curves shown in Fig. 4 comply well with the

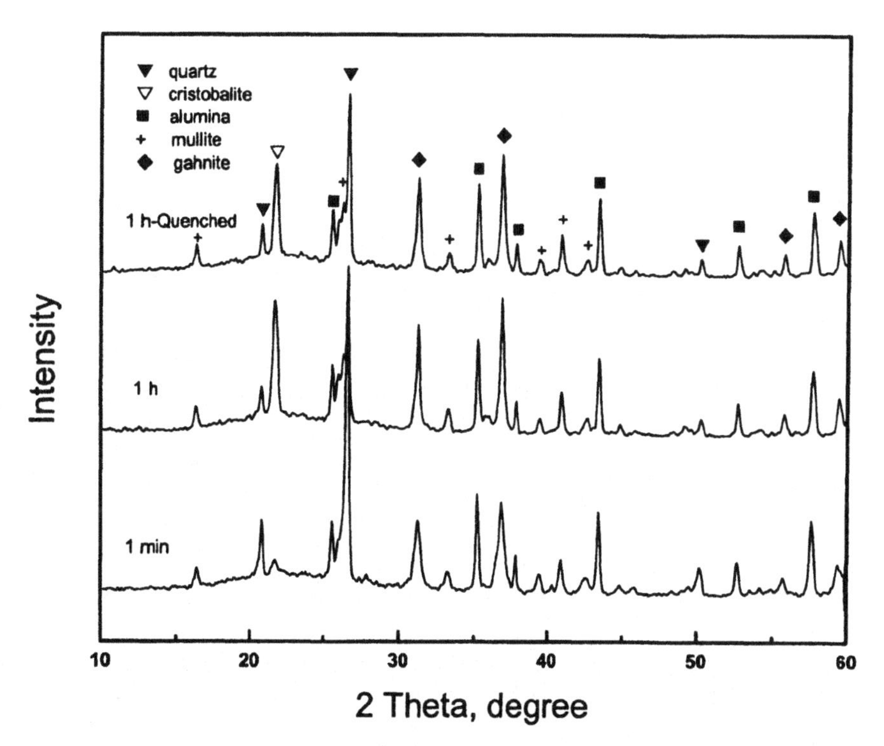

Figure 2. X-ray diffraction patterns of specimens with 5 wt% ZnO furnace-cooled after 1 min and 1 h of isothermal holding, and water quenched after 1 h of isothermal holding at 1250°C.

results of diffraction analysis. For the specimens with ZnO, a broad exothermal peak was observed from 1130°C, in contrast to there being no such peak for the reference specimen. On the other hand, mullite began to form at somewhat lower temperature — around 1100°C — and quartz gradually dissolved above 1150°C with the formation of other crystalline phases. Similar to the results shown in Fig. 1, the diffraction intensities of alumina did not change, implying again that the alumina did not participate in the reactions and remained mostly intact during sintering.

From X-ray diffraction analyses, gahnite formation is assumed to be related to the ZnO dissolution into glass phase and its crystallization as gahnite, instead of a direct solid-state reaction between ZnO and Al_2O_3-containing raw materials. To check this assumption, the specimen with 5 wt% ZnO, which was sintered at 1300°C for 30 min, was analyzed through SEM and TEM. When the specimen was etched in HF solution, abundant

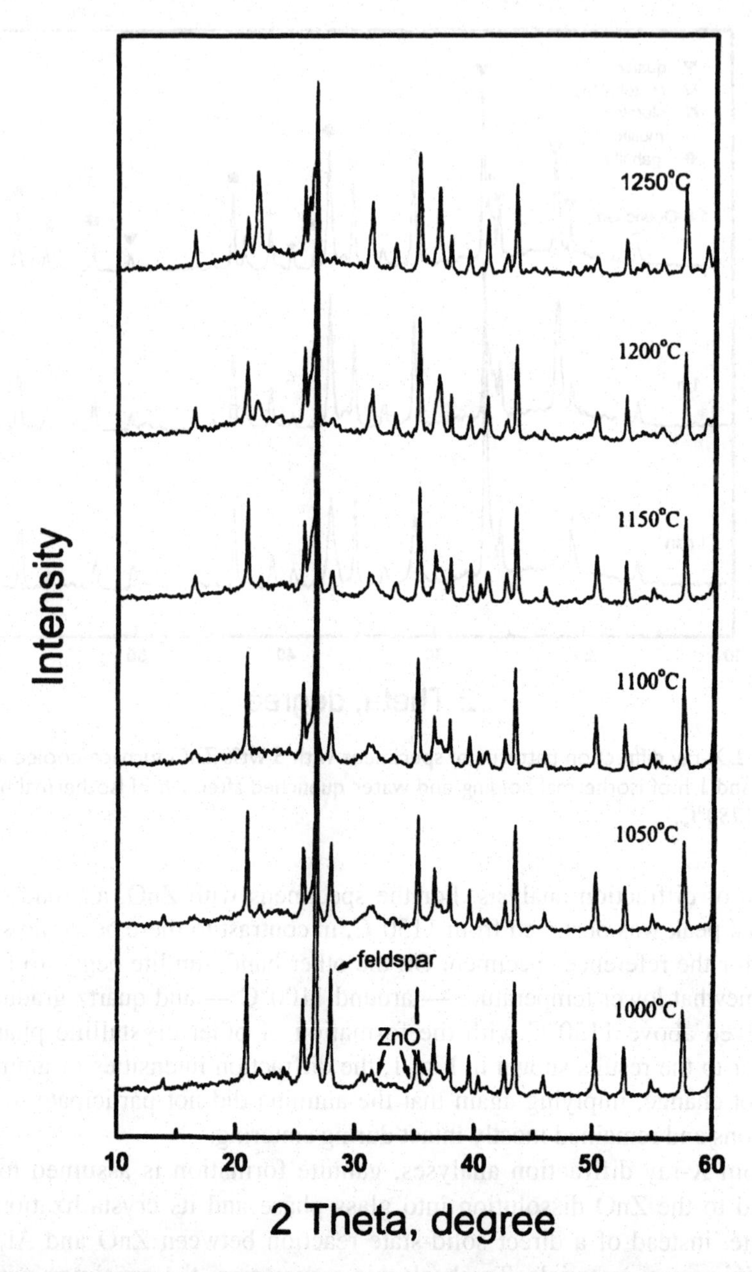

Figure 3. X-ray diffraction patterns of specimens with 3 wt% ZnO sintered at various temperature for 30 min.

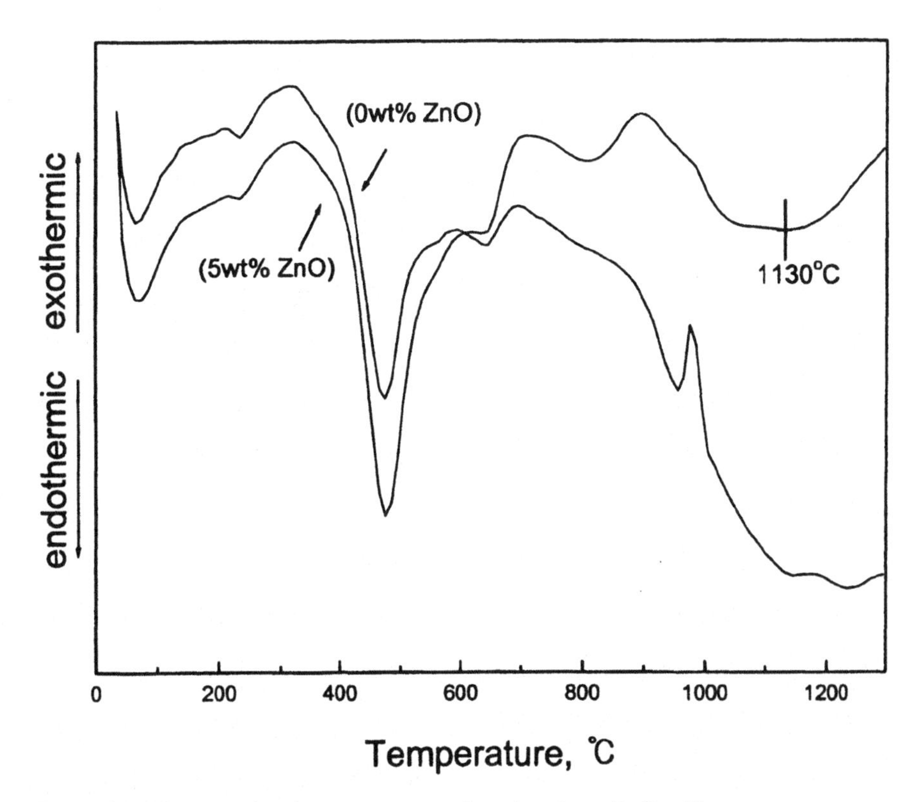

Figure 4. DSC curves for the specimens with and without ZnO addition.

fine crystals in the glass phase were observed. These crystals were also confirmed in SEM and TEM micrographs showing crystal sizes of 50–400 nm [Fig. 5(a,b)]. As shown in Fig. 5(c), EDX analysis on the crystal indicated by an arrow in Fig. 5(b) reveals that the crystal consists of ZnO and Al_2O_3 without any other elements. Since the gahnite is the only binary compound in ZnO-Al_2O_3 phase diagram, the crystal must be gahnite. These experimental results with X-ray diffraction analysis support our assumption that the gahnite crystallized inside the glass phase, which was formed from feldspar melting and ZnO dissolution during heating to sintering temperature.

The gahnite crystallization temperature of 1130°C in the present study is much higher than in other investigations on glass and glass ceramics.[7–12] Escardino et al.[10] observed that gahnite crystallization started from 950°C in the SiO_2-Al_2O_3-ZnO-CaO system, which was preceded by anorthite formation and assumed to occur by decomposition of anorthite. In the present study, however, anorthite is not observed, which might be due to negligible CaO content in the studied specimen, and gahnite crystallized above 1130°C. In spite of glass formation and ZnO dissolution from 1000°C, the gahnite crystallization from a much higher temperature (1130°C) implies that viscosity of ZnO-dissolved glass phase must affect crystallization.

Gahnite formation might be one of the causes promoting the cristobalite crystallization observed in Fig. 1. Consumption of Al by gahnite precipitation in the glass might assist in cristobalite precipitation, although experimental evidence is not sufficient. The mechanism of promoted cristobalite formation in this system is under further investigation.

The water absorption and flexural strength of the sintered specimen were measured. Water absorption decreased with increases in sintering temperature and ZnO content (Fig. 6). Compared to the usual porcelain body, whose water absorption is lower than 0.5%, the sintered specimens have high water absorptions. In spite of these high porosities, the mechanical strengths shown in Fig. 7 are quite high. For the specimen sintered at 1300°C, as ZnO content increased to 5 wt%, flexural strength im-

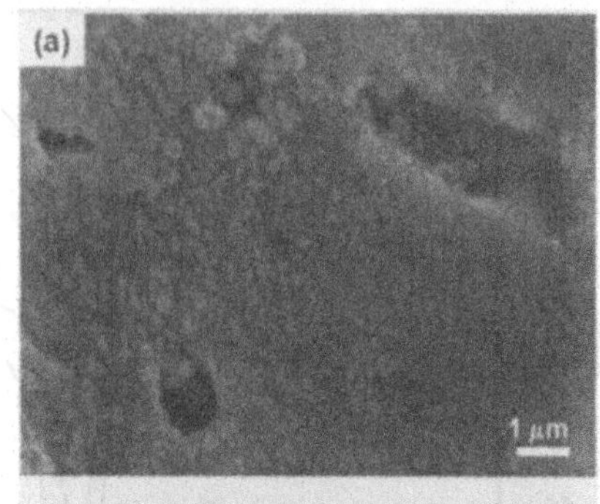

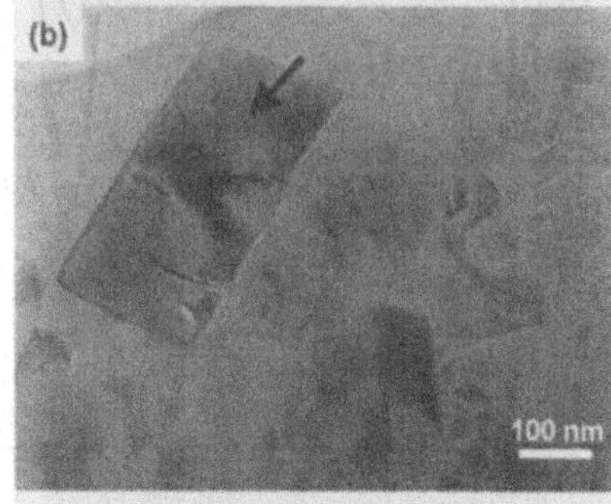

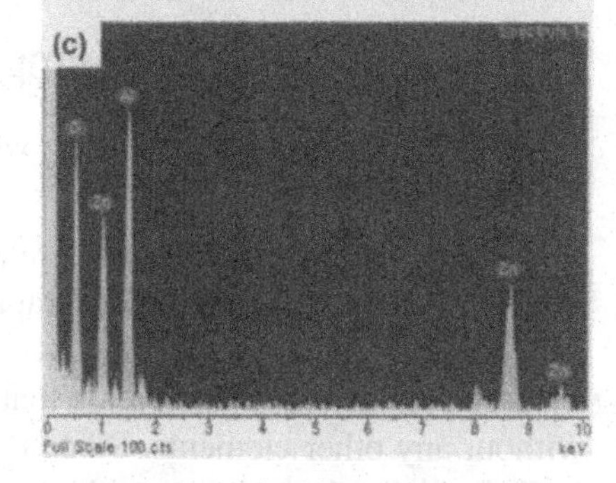

Figure 5. Microstructure and composition analysis of glass region in the specimen with 5 wt% ZnO sintered at 1300°C for 30 min and etched in 10% HF solution for 1 min: (a) SEM micrograph; (b) TEM micrograph; (c) EDX pattern taken from the crystal indicated by an arrow in (b).

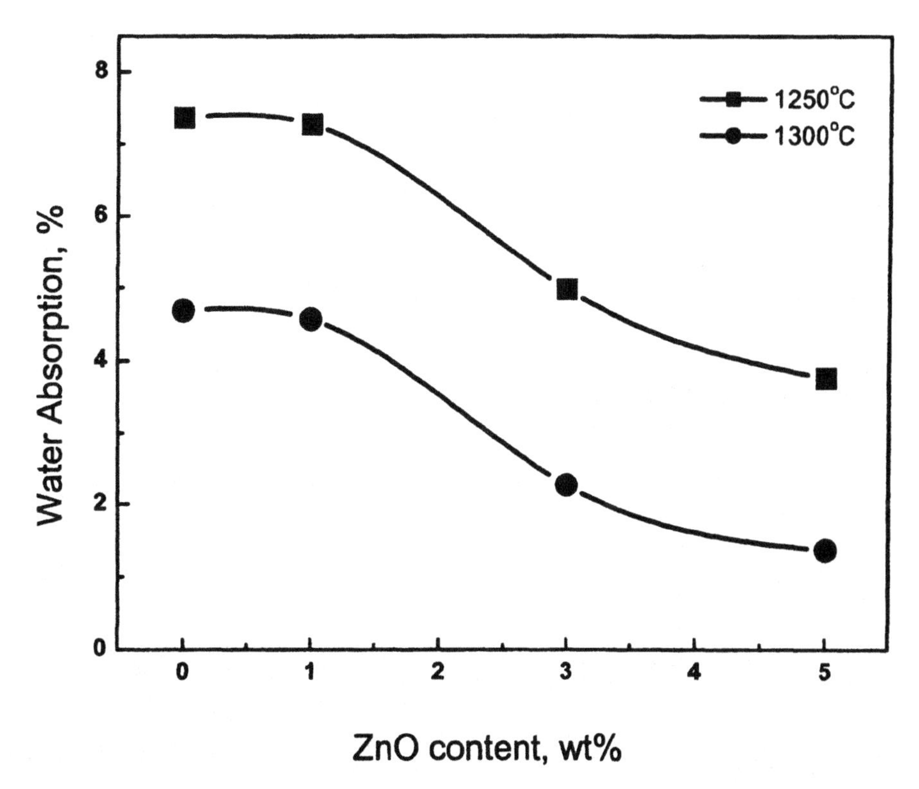

Figure 6. Water absorptions of specimens with ZnO addition sintered at 1250 and 1300°C for 30 min.

proved from 87 to 105 MPa, about a 20% increase. As sintering temperature was lowered to 1250°C, water absorption increased considerably (e.g., from 1.4 to 3.8% for the specimen with 5 wt% ZnO) but strength reduction was moderate: from 106 to 100 MPa. This means that chemical composition and crystallization contributed more than densification to the strength improvement in the investigated sintering temperature range.

Wear resistance of developed porcelain in the present investigation has been measured. Figure 8 shows the results of the weight loss after abrasion test on the sintered specimen. With ZnO content, the weight loss was reduced considerably, revealing improved wear resistance. This might be due to high crystallinity of specimens with ZnO.

This study was intended to impart moderate open porosity but high wear resistance and high flexural strength to porcelain for potential application in

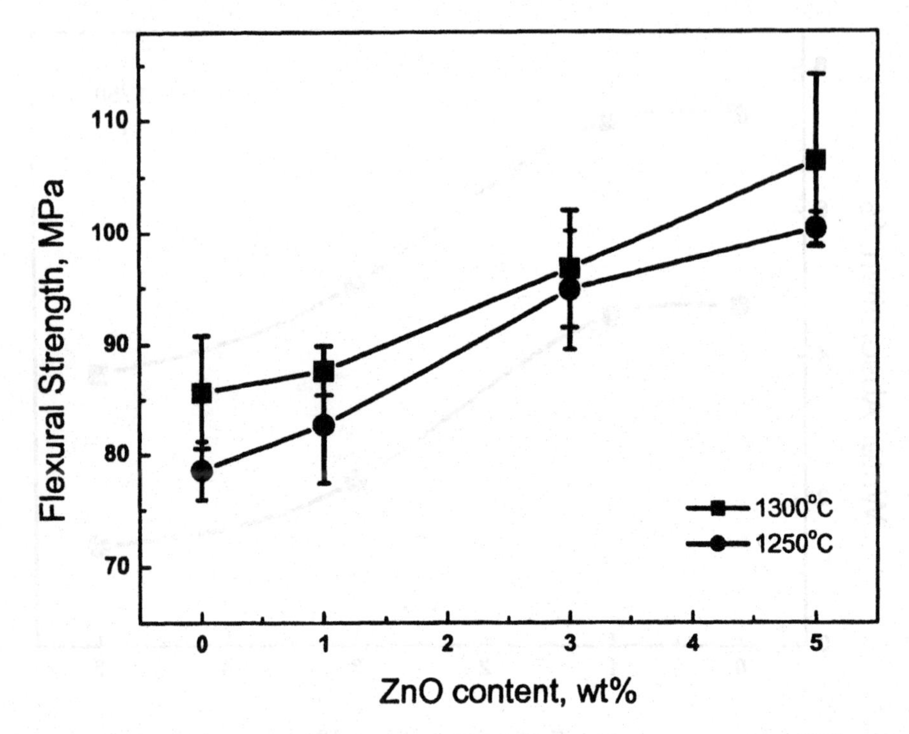

Figure 7. Flexural strengths of specimens with ZnO addition sintered at 1250 and 1300°C for 30 min.

making abrasive parts for food-crushing equipment. Although ceramics have been promising candidate materials for abrasive parts because of their superior wear resistance and better safety compared to cast iron, the low crushing efficiency of dense ceramics due to the low friction coefficient prevented their practical use. The porcelain body with high open porosity developed in the present investigation must have high adherence to crushed materials and result in high crushing efficiency. In addition, since the raw materials used in this study are mostly from natural resources, the abrasive parts could be manufactured economically.

Conclusion

The crystallization behavior of porcelain with ZnO addition has been investigated. As increase in ZnO content, gahnite ($ZnAl_2O_4$) phase was crystallized and cristobalite formation was promoted. However, alumina added as a raw material remained mostly intact during sintering and mullite forma-

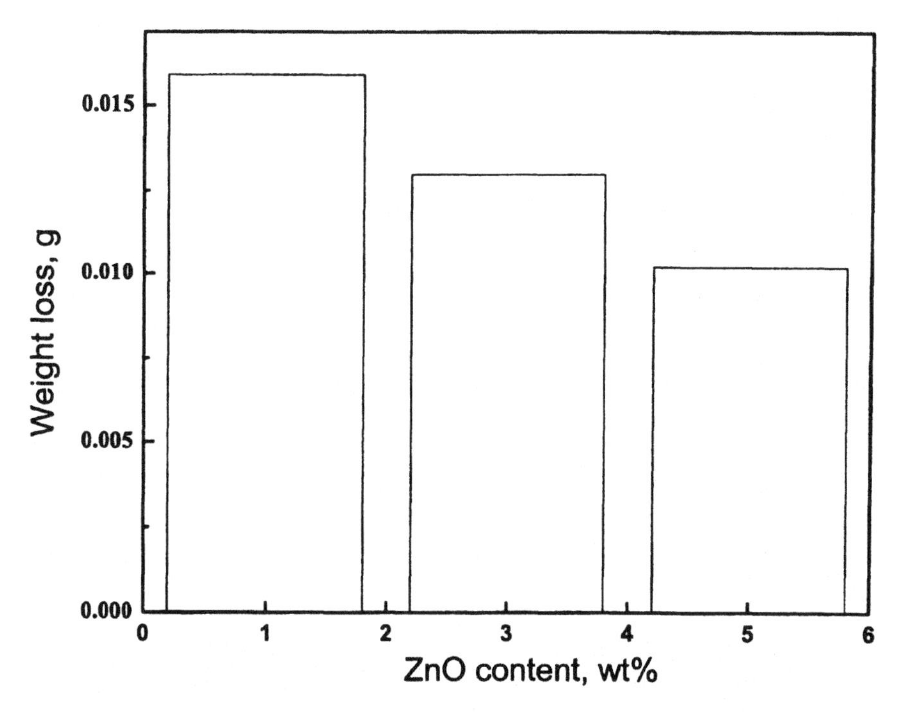

Figure 8. Weight loss in the abrasion test of the specimens with 3 wt% ZnO sintered at 1250°C for 30 min.

tion was not affected by ZnO addition. The added ZnO survived up to 1000°C and disappeared with glass formation due to feldspar melting. The gahnite crystals started to form inside the glass phase from 1130°C when glass was significantly formed and its viscosity was low enough for the gahnite crystallization. The gahnite crystals, with sizes of 50–400 nm, were homogeneously distributed in the glass phase after sintering. These experimental results suggest that gahnite crystallization comes after glass formation due to feldspar melting and ZnO dissolution into the glass.

Water absorption, flexural strength, and wear resistance were measured. With increases in ZnO content and sintering temperature, water absorption decreased and flexural strength improved. The strength depended on ZnO content rather than water absorption. Wear resistance was significantly improved with ZnO. These variations of strength and wear resistance appear to be closely related to gahnite formation due to ZnO addition. The porcelain developed in this study might be a candidate material for abrasive parts in food-crushing equipment, which require moderate porosity and high strength with low manufacturing cost.

References

1. R. A. Eppler and D. R. Eppler, *Glazes and Glass Coatings.* American Ceramic Society, Westerville, Ohio, 2000. Pp. 64–65.
2. C. W. Parmelee, *Ceramic Glazes.* Cahners Publishing Company, Boston, 1973. Pp. 44–47.
3. J. R. Taylor and A. C. Bull, *Ceramic Glaze Technology.* Pergamon Press, 1980. Pp. 35, 41–44.
4. R. A. Eppler and D. R. Eppler, *Glazes and Glass Coatings.* American Ceramic Society, Westerville, Ohio, 2000. Pp. 37–39.
5. F. H. Norton, *Ceramics for the Artist Potter.* Addison-Wesley Publishing Company, 1956. Pp. 247–248.
6. C. W. Parmelee, *Ceramic Glazes.* Cahners Publishing Company, Boston, 1973. Pp. 544–545.
7. B. Karasu, M. Çaki, and S. Turan, "The Development and Characterisation of Zinc Crystal Glazes Used for Amakusa-like Soft Porcelains," *J. Euro. Ceram. Soc.,* **20**, 2225–2231 (2000).
8. B. Karasu, M. Çaki, and Y. G. Yesilbas, "The Effect of Albite Wastes on Glaze Properties and Microstructure of Soft Porcelain Zinc Crystal Glazes," *J. Euro. Ceram. Soc.,* **21**, 1131–1138 (2001).
9. B. Karasu and S. Turan, "Effect of Cobalt, Copper, Manganese, and Titanium Oxide Additions on the Microstructures of Zinc-Containing Soft Porcelain Glazes," *J. Euro. Ceram. Soc.,* **22**, 1447–1455 (2002).
10. A. Escardino, J. L. Amoros, A. Gozalbo, M. J. Orts, and A. Moreno, "Gahnite Devitrification in Ceramic Frits: Mechanism and Process Kinetics," *J Am. Ceram. Soc.,* **83**, 2938–2944 (2000).
11. E. Tkalcec, S. Kurajica, and H. Ivankovic, "Isothermal and Nonisothermal Crystallization Kinetics of Zinc-Aluminosilicate Glasses," *Thermochimica Acta,* **378**, 135–144 (2001).
12. B. E. Yekata and V. K. Marghussian, "Sintering of $\beta.q_{ss}$ and Gahnite Glass Ceramics," *J. Euro. Ceram. Soc.,* **19**, 2963–2968 (1999).
13. W. Holand and G. Beall, *Glass-Ceramic Technology.* American Ceramic Society, Westerville, Ohio, 2002. Pp. 112–115.
14. ASTM C373-88, "Standard Test Method for Water Absorption, Bulk Density, Apparent Porosity, and Apparent Specific Gravity of Fired Whiteware Products." American Society for Testing and Materials, 1999.
15. KS L1001, "Ceramic Tiles." Korean Agency for Technology and Standards, 1997.

Benbow Analysis of Extruded Alumina Pastes

C. R. August and R. A. Haber
Department of Ceramic and Materials Engineering, Rutgers University, Piscataway, New Jersey

Capillary extrudate analysis was performed on two aqueous alumina pastes having average particle sizes of 1 and 5 μm, respectively. The batches were composed of hydroxypropyl methylcellulose binder and a stearate lubricant. Extrusion as a function of solids loading, batch mixedness, and extrusion velocity was measured. Results show that the initial particle size and solids loading influenced the uniformity and appearance of the cylindrical extrudates. Further variation of particle size distribution showed a strong pressure versus length/diameter relationship.

Introduction

The term "extrusion" refers to the forcing of a paste through a specified shape or die. This procedure is used in many industrial applications for production of a wide array of manufactured goods. Extrusion is used extensively in the food industry, and is also used in ceramic applications such as the production of catalytic converters and superconductor coils. Much research is still needed in the area of extrusion in order to fully understand the flow properties of a system.

Capillary rheometry is the constant rate extrusion of a material through a specific die land. This testing method gives apparent viscosity versus shear rate data, which can then be used to determine the behavior of the various stresses on the system. Capillary rheometry is very useful in analyzing the flow properties of pastes and can be used to help model the flow properties of complicated systems.

There are a few methods of using capillary rheometry to analyze a paste. The Bagley method[1] involves using dies of cross slit and round geometries in order to evaluate the effects of die geometry on extrusion pressure. Benbow analysis[2] uses dies of varying length over diameter at a series of extrusion speeds to calculate certain constants. These constants can be related to properties of the initial paste. This method can then be used to predict extrusion behavior of a paste in more complicated systems.

The Benbow parameters calculated using capillary rheometry data refer to such properties as the bulk stress and the yield stress of the material flowing through the die land. Benbow analysis also calculates two parameters that relate to velocity of the paste in relation to the barrel wall and die land. These parameters are calculated using the Benbow equation:

$$P = 2(\sigma_0 + \alpha V^m)\ln(D/D_0) + 4(\tau_0 + \beta V^n)(L/D)$$

Where P is extrusion pressure, V^m is velocity of the paste in the die land, D is diameter of the extrusion die, D_0 is diameter of the capillary barrel, V^n is velocity of the paste near the wall, and L is length of the extrusion die. The parameters calculated from this equation are σ_0 (bulk stress), τ_0 (yield stress), α (velocity factor, die land) and β (velocity factor, barrel wall).

It is important to understand the geometry of the capillary rheometer to fully comprehend parameters in the Benbow equation. Figure 1 is a diagram of the capillary rheometry barrel with the parameters needed for the Benbow equation labeled.

Experimental Procedure

Batch Composition

Four basic components were used to create an extrudable paste: powder, liquid, binding agent, and lubricating additive. In this experiment the batches used Alcoa alumina as the powder, distilled water as the liquid phase, an Aqualon methyl celluloid binder, and a Fisher Scientific sodium stearate lubricant. The batch compositions are listed in Table I.

The procedure for making a batch for capillary rheometry is as follows. The standard batch formulation was 96% alumina (single alumina or alumina blend, 3% binder (HPMC), and 1% stearate-based lubricant additive. The dry components of the batch were weighed out in 1000 g batch and mixed in a ball mill for a 1 h. The powder was separated from the media using a dry sieving technique. This powder was then mixed with the liquid component in a dual sigma blade mixer for about 30 min, at which point the material was considered plasticized. The batch was then double extruded through a 0.5 in. diameter die to ensure homogeneity.

Capillary Rheometry

The capillary rheometer was assembled on an Instron 4500 Series (Fig. 2). This equipment consisted of a capillary barrel, pressure meter, transducer, plunger, ram, die head, dies, and support structure. The barrel was screwed into the support structure, which was fixed to the crosshead of the Instron.

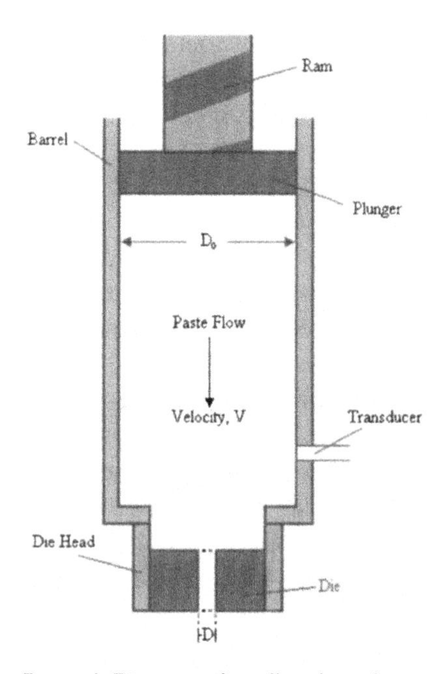

Figure 1. Diagram of capillary barrel.

Figure 2. Capillary rheometer and Instron 4500.

Table 1. Batch compositions

Batch	Alumina	Water (%)	Lubricant
1	A2	19	1.0% sodium stearate
2	A10	19	1.0% sodium stearate
3	A2/A10	19	1.5% sodium stearate
4	A2/A10	19	2.0% sodium stearate

The transducer was fixed into the extrusion barrel, the hole located behind the die, and connected with the power meter. The die was placed in the die-head assembly, which was then threaded into the bottom of the barrel. The Instron was calibrated so that the crosshead could be moved. With the crosshead all the way down, a test batch was placed into the barrel. The plunger was then set on top of the material and the ram fixed to the load cell suspended from the top of the Instron. The crosshead was then raised manually until the ram was just above (without touching) the plunger. This was achieved by visually observing the power meter as the crosshead was raised. When the pressure spiked, the ram had made contact with the plunger and the crosshead was lowered until the pressure stabilized, indicating that the ram was no longer in contact with the plunger. The power meter was then tared.

At this point, the time-based pressure readings began. The speed of the crosshead programmed into the method was calculated for the desired extrusion velocities of 2.00, 0.20 and 0.02 in. of extrudate per second. Since the speed of the crosshead was directly related to the speed at which the extrudate flowed from the die, the calculated factor need only be adjusted by powers of 10 to attain the other required testing speeds.

Once the computer-designed testing program commenced, the crosshead would rise at the desired speed. The pressure indicated on the power meter would begin to spike and could go up to very high pressures (in the range of 10 Ksi) in an effort to overcome the static yield stress required to initially force the batch through the 50 mil diameter die. Once this yield stress was attained and the material began to flow through the die, the pressure would begin to decrease. The pressure was monitored over this decline until it leveled off. This value was then recorded. Once the pressure had been obtained at all speeds for all dies involved, the die head was removed and the crosshead was manually raised to push out the remaining material and the plunger.

Benbow Analysis

Once capillary rheometry had been performed, a Benbow analysis could be conducted. The calculation of the Benbow parameters came directly from the capillary rheometry data. Each batch was extruded through at least two dies at two different speeds in order to have enough information to calculate these parameters. To begin the calculations, a best-fit line was applied to the capillary data. Figure 3 is an example of capillary data with values needed for the Benbow calculations labeled. The equations that define the Benbow parameters are as follows.

$$\sigma_o = \frac{(OA)V_2 - (OB)V_1}{2(V_2 - V_1)\ln(D_o/D)}$$

$$\alpha = \frac{(OB) - (OA)}{2(V_2 - V_1)\ln(D_o/D)}$$

$$\tau_o = \frac{V_2[(CJ) - (OA)] - V_1[(DJ) - (OB)]}{4(OJ)(V_2 - V_1)}$$

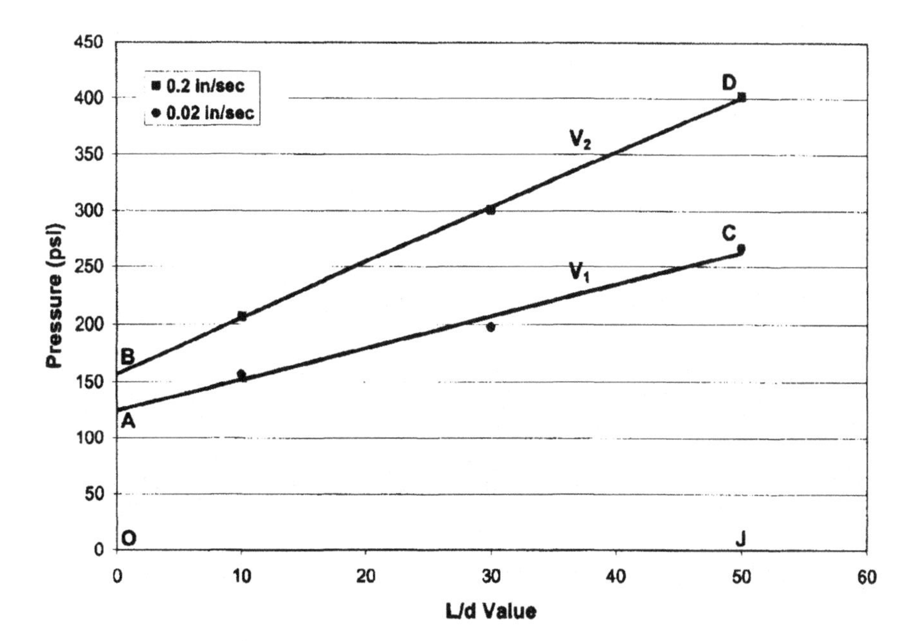

Figure 3. Values required for Benbow analysis.

$$\beta = \frac{[(DJ)-(OB)]-[(CJ-(OA)]}{4(OJ)(V_2-V_1)}$$

These parameters were calculated for the all the batches being examined.

Results and Discussion

Batch compositions for this experiment were chosen so that certain variables could be examined. Two different Alcoa aluminas were used so that the effect of particle size distribution could be explored. Figure 4 shows a comparison of the particle size distributions of the A2 and A10 aluminas. The percentage of Fisher Scientific sodium stearate added to the composition was also varied. Capillary rheometry for Batch 1 at three speeds and for three different dies can be seen in Fig. 5. The Benbow parameters calculated for each batch can be seen in Table II.

The Benbow constants calculated from the capillary rheometry data can be grouped into two sections. Parameters α and σ_0 are calculated from the behavior of the flow in the barrel. Parameters β and τ_0 are calculated from the behavior of flow in the die. Variables α and β are velocity-dependent fac-

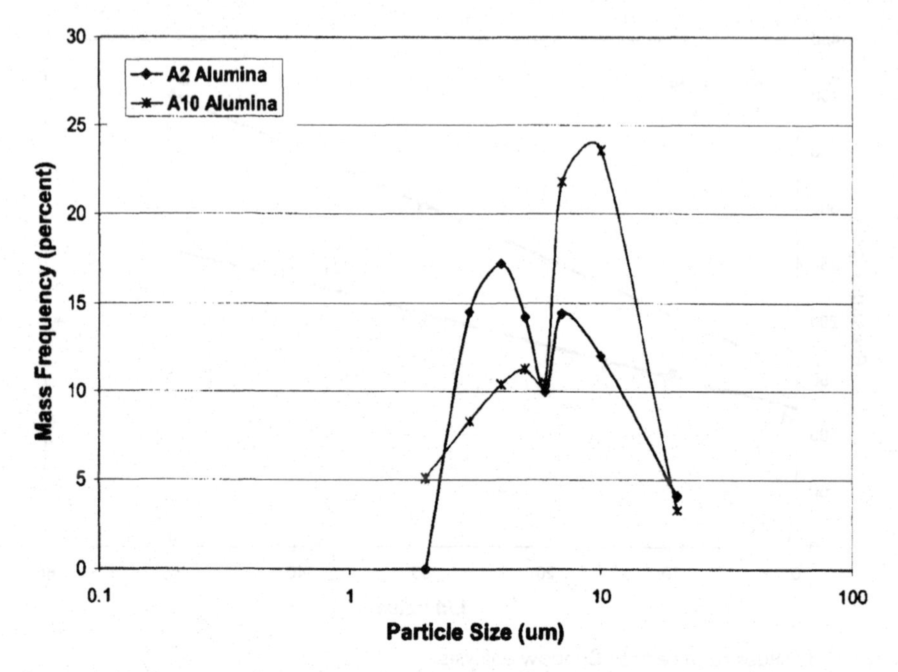

Figure 4. Alumina particle size distribution comparison.

tors. These factors have been shown to relate to the viscosity of the material.

We can use the calculated Benbow parameters to compare different paste formulations. Looking at the τ_o for Batch 1 and Batch 2, we can see that the yield stress parameter is greater for the batch containing the finer alumina. This makes sense since the finer alumina would have a greater number of particles in the paste. This greater number of particles would mean there are more particle contacts to break in order for flow to occur.

The increased addition of sodium stearate lubricant also has an effect on the Benbow parameters. Capillary rheometry data showed that greater pressure was required to extrude the 2.0% sodium stearate paste. The Benbow parameters σ_o and τ_o both increase from Batch 3 to Batch 4. This implies that an increase in the percentage of sodium stearate will increase the bulk stress as well as the yield stress.

Conclusion

Analysis of four alumina batches with capillary rheometry and Benbow calculations shows that coarser particle size distributions have lower yield stress to overcome. These coarser pastes also have greater bulk stress,

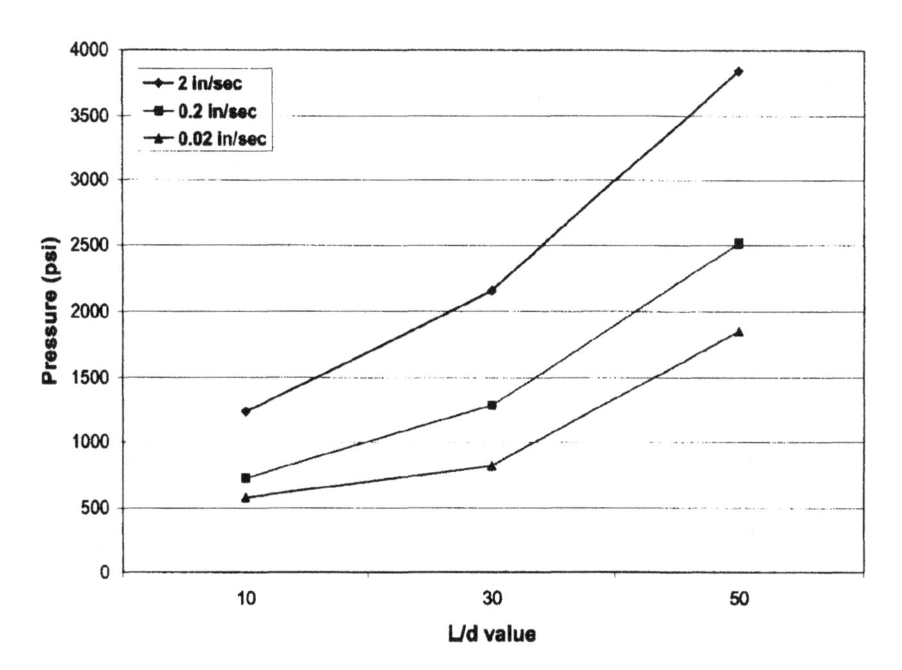

Figure 5. Capillary rheometry data for Batch 1.

Table II. Benbow parameters

Batch	σ_o (psi)	α (psi s/in.)	τ_o (psi)	β (psi s/in.)
1	7.63	71.78	7.81	19.59
2	9.9	30.28	5.35	12.06
3	43.99	73.46	2.5	19.33
4	47.77	52.33	3.38	22.71

which can translate to greater green strength of the extrudate. The addition of sodium stearate to a batch can raise extrusion pressures, but this increase in pressure is due to an increase in the bulk stress of the material. The additives may therefore be beneficial enough to the green strength of the material to warrant extruding at a higher pressure.

References

1. E. B. Bagley, *J. Applied Physics* **28** (1957), 624–627.
2. J. Benbow and J. Bridgwater, *Paste Flow and Extrusion.* Clarendon Press, Oxford, 1993.

Microscopy Methods for Ceramic Applications

C. Collins and E. Westbrook
Unimin Corporation, Tennessee

A brief examination of several methods of applied microscopic tools for the evaluation of properties and performance of ceramic materials, processing, and finished products is presented. Examples of optical and scanning electron microscopic investigations will also incorporate the use of energy dispersive spectroscopy (EDS) and various methods of sample preparation.

Pyroplastic Deformation Revisited

Aubree M. Buchtel and William M. Carty
Whiteware Research Center, New York State College of Ceramics at Alfred University, Alfred, New York

Mark D. Noirot
U.S. Borax Inc., Valencia, California

A simple, robust test method was developed for measuring pyroplastic deformation of a whiteware body based upon the calculated stress on a body during firing. Stress is determined from sample geometry, size, and span. A pyroplastic index is generated based on sample stress and deformation, which successfully demonstrates the ability to characterize the deformation tendency. Two deformation events are apparent in deformation versus firing temperature data, both due to viscous flow. A significant portion of deformation occurs in a non-steady-state region of firing, far below the dwell temperature, proposed to be due to heterogeneities caused by microstructural evolution of the body. A secondary deformation event, proposed to be steady-state creep, occurs at the dwell temperature. Quartz dissolution ceases between 1 and 2 h of dwell time at the peak temperature, and the amount of mullite remains constant, therefore the microstructure is proposed to be steady-state. Through the measurement of bending sample deformation, strain and strain rate can be calculated. The steady-state creep equation can then be used to calculate the system viscosity.

Background

Pyroplastic deformation traditionally has been measured using many different techniques. Several methods have been proposed to determine a pyroplastic index, allowing the comparison of different body compositions or firing curves. The most common index used in literature was proposed by McDowall and Vose,[1] who base their assumptions on a three-point bending stress on a sample supported at two ends:

$$\frac{S}{t} = \frac{5\rho g L^4}{72\eta d^2} \qquad (1)$$

Rearranging the equation allows for the definition of an index using the dimensions of a sample.

Microstructural Evolution

Understanding the microstructural evolution of a whiteware body is essential to comprehending deformation. Seymour researched the microstructural

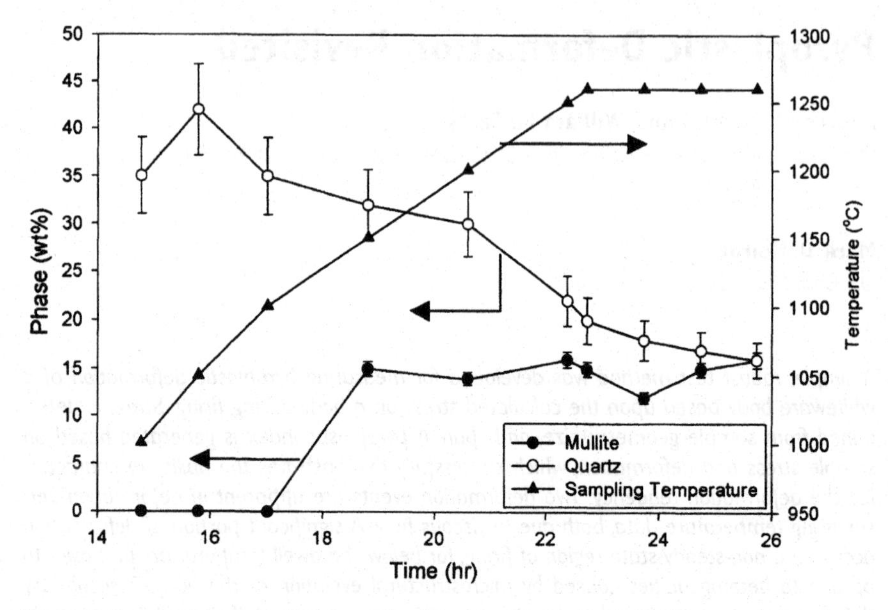

Figure 1. Phase evolution of mullite and quartz as a function of temperature.[2]

evolution of a whiteware body using X-ray diffraction techniques.[2] As shown in Fig. 1, mullite crystallization occurs between 1100 and 1150°C and remains constant above 1150°C. The amount of alumina dissolved into the glass phase is dictated by the amount of alkali in the system. Because the mullite level is constant above 1150°C, the alumina in the glass phase must also remain constant. Quartz dissolution begins around 1100°C. Only increasing temperature will allow more quartz to dissolve. At 990°C (the eutectic), a glassy phase begins to develop around the feldspar particles. Alkali are diffusing out of the particles, creating a low-viscosity glass region in the vicinity of the feldspar particles. Only time will allow the homogenization of the glass phase by diffusion of the alkali throughout the system. It is proposed that these low-viscosity glass phase regions are the primary method by which pyroplastic deformation occurs.

Calculation of Stress

The stress on a specimen is calculated using theory of elasticity. The benefit of calculating stress is that it is normalized to sample geometry and size, therefore, results can be compared for different samples. Due to density and dimension changes during densification, green and fired values were investigated for the most reliable stress level. The stress is calculated using the fired density, geometry and dimensions of the sample, span length, and gravity.

To determine the stress on a sample using the theory of elasticity, several assumptions are made. The first assumption is that shear stresses are negligible. Shear stresses are not often of controlling importance except for small span-to-depth ratios.[3] If the length of the sample is at least five times the height, shear forces are half those of the bending stress for an I-beam.[4] A similar calculation for a rectangular bar demonstrates that the smallest ratio of bending stress to shear stress used in pyroplastic deformation presented herein is 9:1.[5] These two statements support the idea that shear forces can be neglected. The second assumption is that the system is perfectly elastic.[6] The whiteware system is certainly not perfectly elastic; however, it is shown later that increasing stress results in a linear trend of increasing deformation until a maximum stress where the system becomes nonlinear. Therefore it is assumed that the material obeys Hooke's law within the linear region, and the elastic analogy can be assumed. A similar assumption is that deformation is within the proportional limit. This is the greatest stress a material can sustain without deviating from the law of stress-strain proportionality.[3] Because of the linear relationship between stress and deformation in a particular region, it can be assumed that the system is obeying Hooke's law. Lastly, it is assumed that the material is homogenous and isotropic. Assuming the system has randomly distributed particles, when examining a part of it that is large compared to the constituents in it, the material can be assumed to be isotropic.[6]

The bending stress (σ_B) for samples is

$$\sigma_B = \frac{Mc}{I} \tag{2}$$

where $M \equiv$ the resultant internal moment in sample, $c \equiv$ the perpendicular distance from the neutral axis to a point farthest away from the neutral axis where the maximum stress occurs, and $I \equiv$ the moment of inertia of the cross-sectional area computed about the neutral axis.[7]

The internal moment of the samples is calculated using a distributed load dictated by force due to gravity and sample mass acting uniformly along the sample between supports as in Fig. 2. The internal moment of a sample with a distributed load may be calculated using[8]

$$M = \frac{wL^2}{8} \tag{3}$$

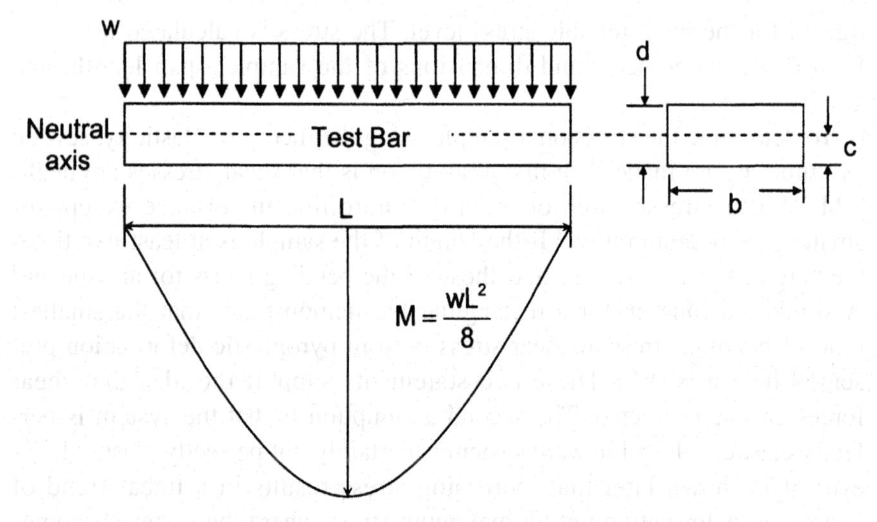

Figure 2. Illustration of internal moment and unit load on a rectangular sample with a distributed load.

where $w \equiv$ unit load (Newton/m) and $L \equiv$ span length between supports (m). Since w is determined by the load on the sample, it will change with sample geometry. The unit load (w) for a rectangular bar is

$$w = \frac{\rho g d b L}{L} \tag{4}$$

and for a circular rod

$$w = \frac{\rho g L \pi r^2}{L} \tag{5}$$

The moment of inertia for a rectangular bar is

$$I = \frac{bd^3}{12} \tag{6}$$

and for a circular rod

$$I = \frac{\pi r^4}{4} \tag{7}$$

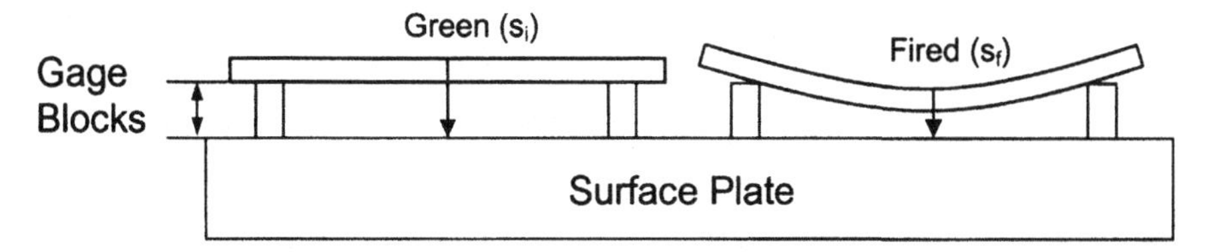

Figure 3. Green and fired deformation measurement.

The reduced equation for the for a rectangular bar is

$$\sigma_B = \frac{Mc}{I} = \frac{wL^2c}{8\cdot I} = \frac{\rho gbdL^3d\cdot 12}{L\cdot 16\cdot bd^3} = \frac{3\rho gL^2}{4d} \quad (8)$$

The reduced equation for the for a circular rod is

$$\sigma_B = \frac{Mc}{I} = \frac{wL^2c}{8\cdot I} = \frac{\rho g\pi r^2L^3\cdot 4}{8L\pi r^4} = \frac{3\rho gL^2}{4d} \quad (9)$$

Deformation Measurement

Deformation is measured prior to and after firing to minimize measurement error. Samples are placed on gage blocks, and a height gage is used to measure the distance from the surface plate to the top of the sample as shown in Fig. 3. The same setup is used for a fired measurement. Because of the nature of the deformation measurement, linear shrinkage must be taken into account when calculating deformation. Therefore:

$$s = s_i - s_f - (\Delta d) \quad (10)$$

$$(\Delta d) = d_i - d_f \quad (11)$$

where $s \equiv$ pyroplastic deformation, $s_i \equiv$ green deformation, $s_f \equiv$ fired deformation, $\Delta d \equiv$ linear shrinkage, $d_i \equiv$ initial thickness, and $d_f \equiv$ final thickness.

Stress versus Deformation

Figure 4 displays the relationship between stress and deformation. Several different sample sizes and geometry noted in Table I were fired at a rate of 10 K/min up to 1290°C with a no dwell. Body 4 listed in Table II was used

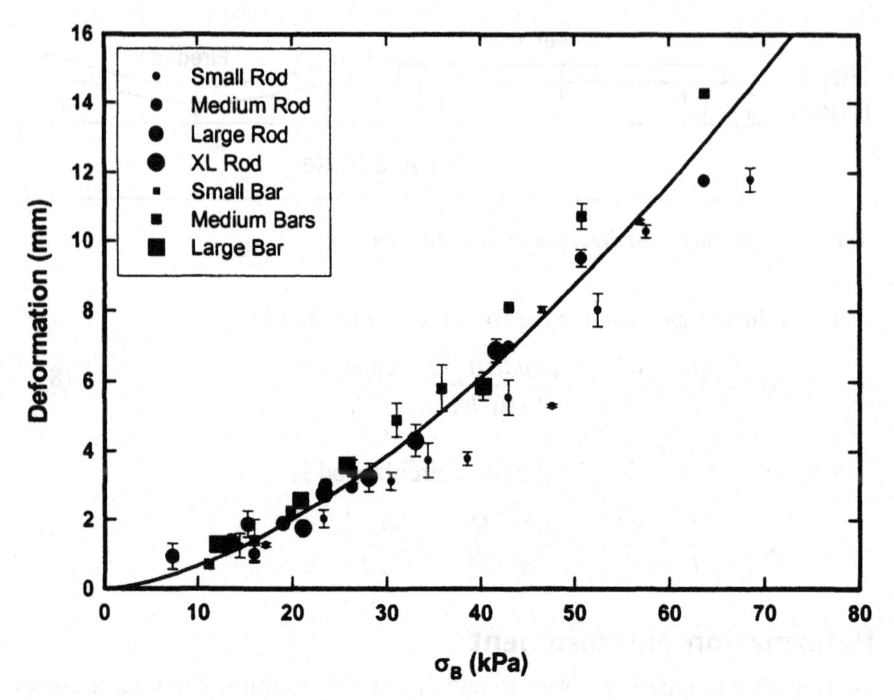

Figure 4. Stress versus deformation of samples with different geometries and cross-sectional areas ($R^2 = 0.94$).

for this experiment. The trend of stress versus deformation is consistent regardless of sample geometry and size.

Effect of Densification on Stress

To determine the effect of densification on stress, green and fired densities were measured for a series of samples. Apparent bulk density of the fired samples was measured following ASTM standard C830-00.[9] For green density measurements, kerosene was substituted for water to prevent sample damage.

Figure 5 displays the results of density and porosity as a function of peak temperature. Samples of Body 4 were fired at a heating rate of 3 K/min to the peak temperature. Apparent density increases and apparent porosity decreases with increasing temperature because of the densification of the body. Figure 6 demonstrates the effect of densification and shrinkage on σ_B. Calculating σ_B as a function of temperature using fired values of sample

Table I. Sample dimensions

Sample geometry	Size	Breadth (mm)	Depth (mm)	Radius (mm)
Bar	Small	10.6	5.50	
Bar	Medium	16.6	8.40	
Bar	Large	22.0	11.0	
Rod	Small			4.00
Rod	Medium			5.50
Rod	Large			7.60
Rod	Extra large			8.50

Table II. Batch composition used for each body (wt%)

Raw material	Body 1	Body 2	Body 3	Body 4
Clay				
TK6	29			
Todd Light	6	10.94	9.95	
Diamond China		18.9	10.45	
M&D		3.98	5.47	3.38
M-23		9.95	9.45	
Old Mine #4				13.13
Kingsley-Rogers				10.00
Excelsior				17.33
Flux				
A 400	21			
G-200		34.81	29.84	31.00
Filler				
Alcan C-71	10			
Sil-Co-Sil 63	34			25.16
Alcan C-70FG			34.32	
Alcoa Alumina				
Flint		20.89		
Plasticizer				
Bentonite Vol Clay		0.53		

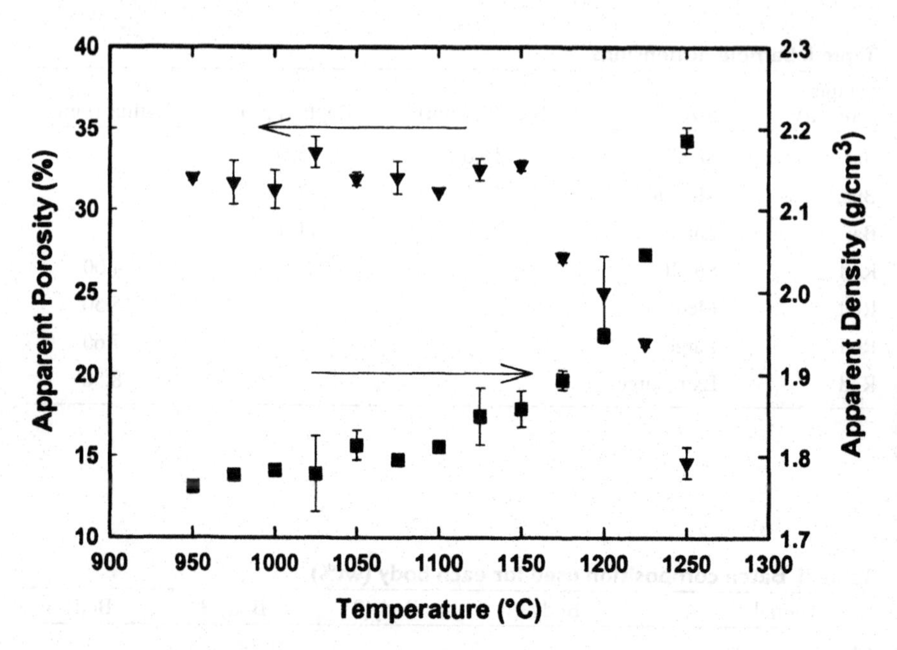

Figure 5. Apparent porosity and apparent density as a function of firing temperature.

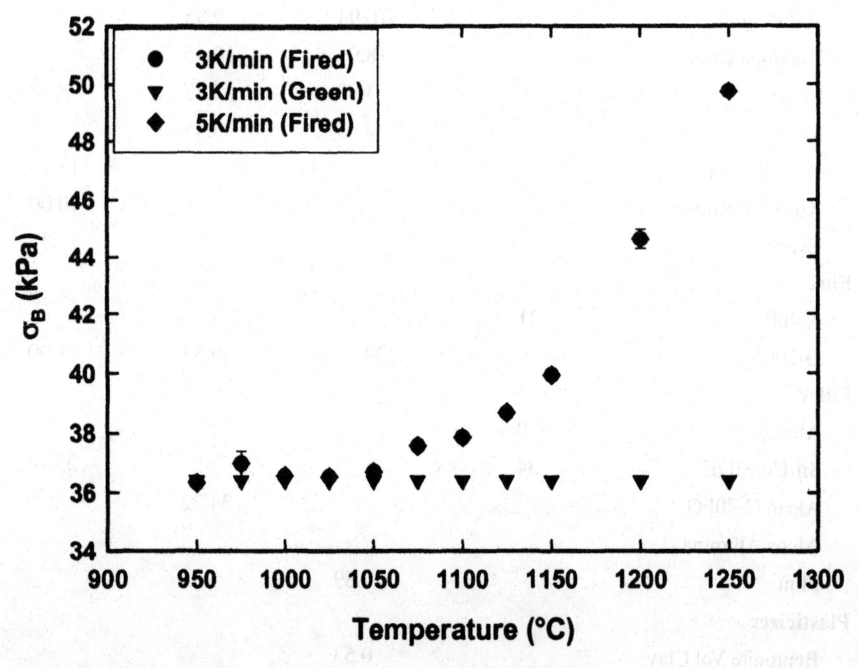

Figure 6. Stress level as a function of temperature. Fired data (3 and 5 K/min) overlay each other, therefore there is no significance of heating rate.

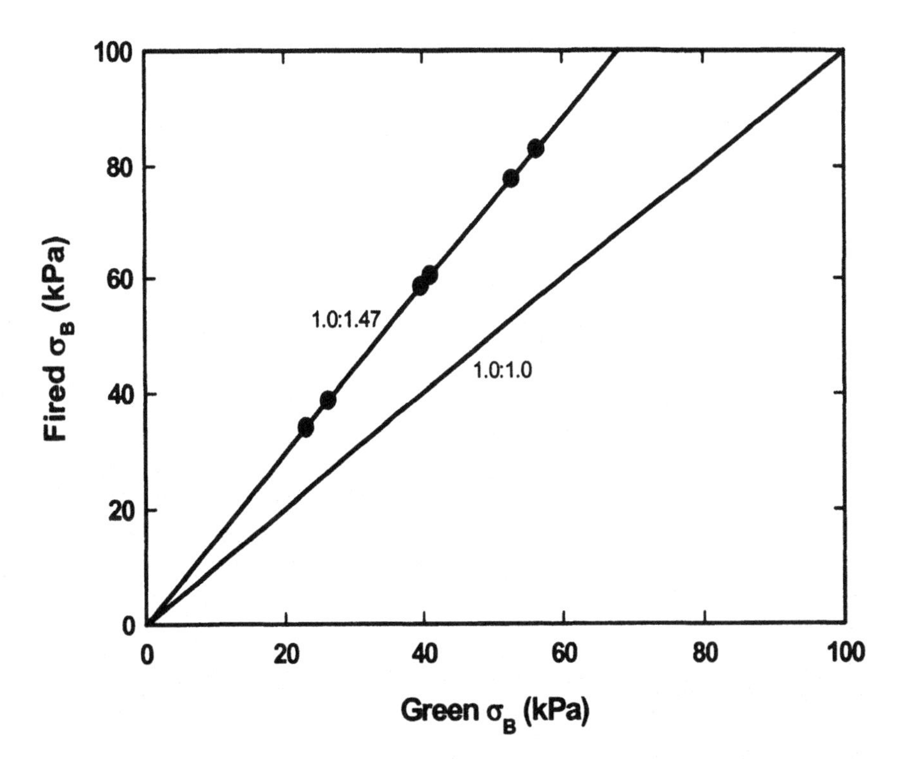

Figure 7. Calculation of stress using green and fired properties establishes a ratio that can be used to estimate the fired stress.

density and height illustrates that stress increases significantly above 1075°C. Figure 7 displays the ratio of fired to green σ_B. Using green sample values requires a correction factor stemming from the densification and shrinkage of the body during firing. Initially, green dimensions of the body can be used to estimate stress. After firing, the stress level can be corrected to reflect the actual stresses during the firing process.

Effect of Overhang

To determine the effect of overhang on deformation, samples of Body 4 were fired with overhang levels between 15 and 45% of the span length, as illustrated in Fig. 8. To identify the span length at which the samples were fired, they were marked at the location where they were touching the setter after firing. Figure 9 establishes the result of increasing overhang length of a sample. Deformation in the center of the sample is reduced with increas-

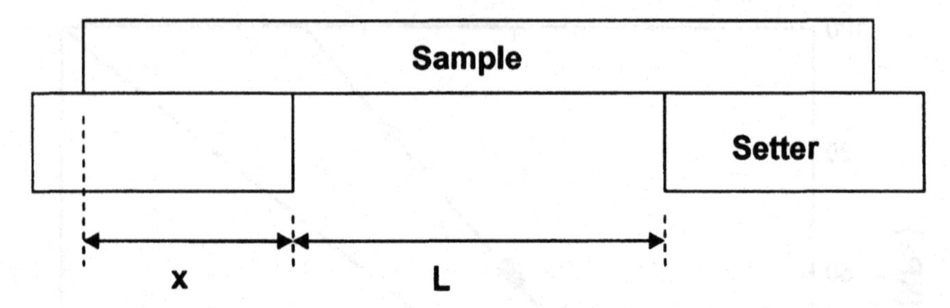

Figure 8. Illustration of sample overhang: % overhang = $[x / (x + L)] \times 100$.

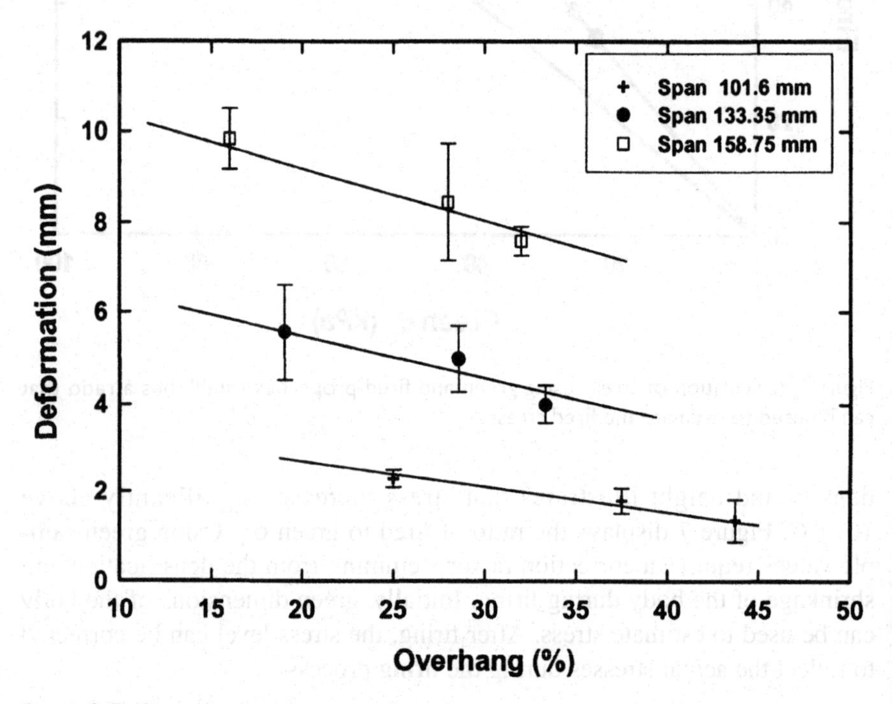

Figure 9. Deformation as a function of overhang.

ing overhang due to introducing an additional stress at each end of the sample. As the overhang increases, the ends of the sample deform, which causes a countermovement and stress. This countereffect results in less deformation in the center of the sample. As a result of the effect of overhang on deformation, all samples were consistently fired with the same overhang amount, which was 15% of the span at each end of the sample. This is the

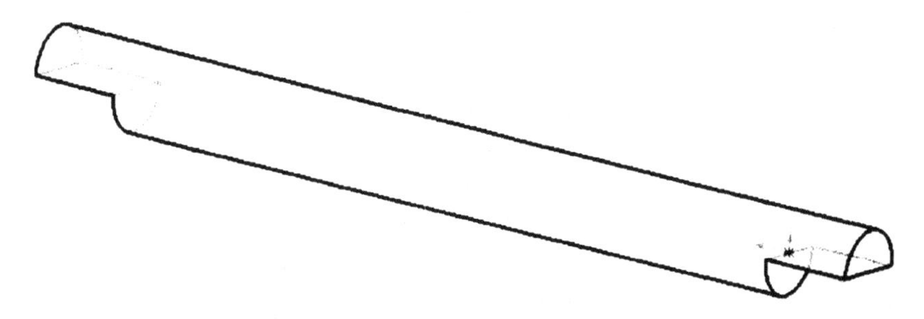

Figure 10. Illustration of flattened rod used to simulate surface area contact between sample and setter of a rectangular bar.

smallest overhang amount in which samples would remain on the setter after shrinkage and deformation take place.

Effect of Surface Area Contact between Sample and Setter

The difference between a rod with a small contact point on the setter and a bar with a large contact area was a possible contribution for inconsistencies in the data when rods and bars were compared. Body 4 was extruded into rods and fired in a tunnel kiln on cast setters. Rods from the same batch were extruded and flattened at the ends as illustrated in Fig. 10. Flattening the ends simulated the contact area rectangular bars would have against the setters.

The results of deformation versus stress level of the experiment are shown in Fig. 11. If there is a difference in contact area that in turn causes the deformation to differ, it is within experimental error (all deformation presented is plus or minus two standard deviations). The data for the flattened rods and the normal rods overlay each other.

Proposed Pyroplastic Index

Initially, a best-fit equation for different body compositions was used to define a constant as the pyroplastic index. As the best-fit equation was being established, the effects of several factors were weighed. The stress region in which theory of elasticity constraints are valid was taken into consideration, as well as scatter in the data and how that scatter affected the best-fit equation. Finally, statistical methods were used to prevent overfitting of the data.[5]

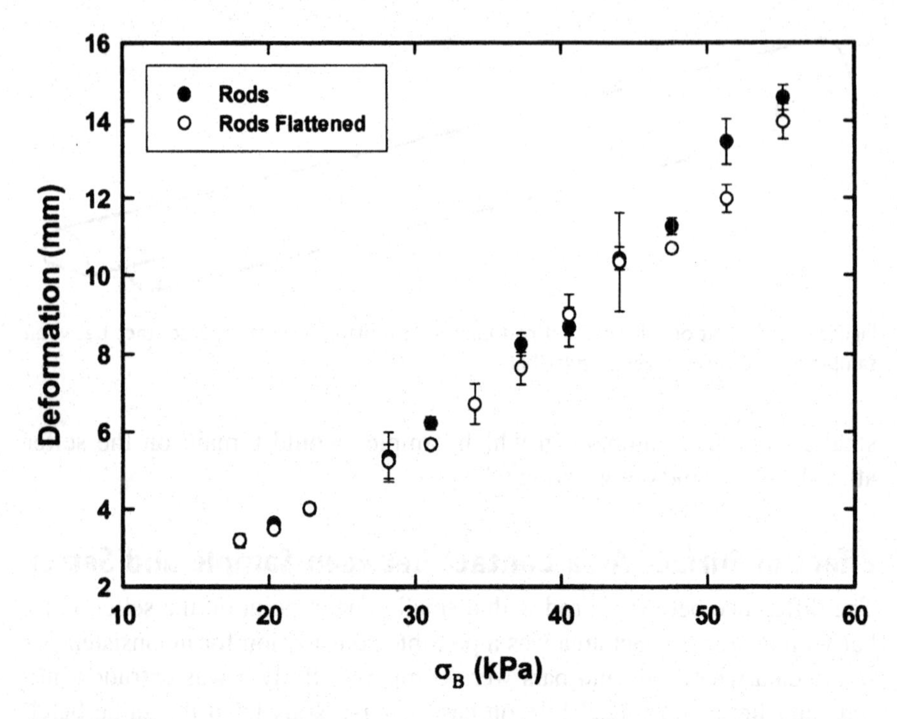

Figure 11. The contact area between sample and setter does not significantly affect deformation.

The resulting index was proposed based on a linear equation, however the constants in the equation were not defined as the pyroplastic indices. The constants were not representative of the deformation tendency of the body. Therefore, a linear regression fit to the deformation versus stress trend allows interpolation of deformation at 50 kPa. The pyroplastic index is proposed to be the deformation at a stress of 50 kPa. The index is efficient and allows different body compositions and processing or firing conditions to be compared.

Figure 12 illustrates four industrial body compositions (listed in Table II) in which the pyroplastic index has been calculated using a linear equation. They pyroplastic deformation was plotted with stress and a linear regression to fit the data. The pyroplastic indices are labeled in the figure legend. The three vertical dotted lines represent what stress levels at which pyroplastic deformation were measured. The solid vertical line represents the stress level at which the pyroplastic index was interpolated.

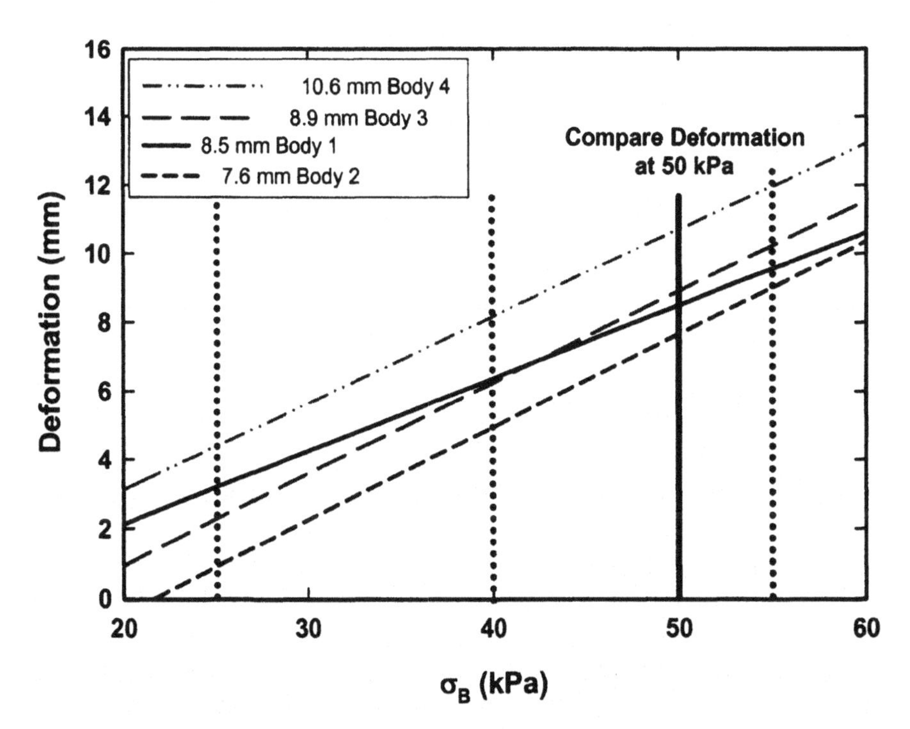

Figure 12. Vertical dotted lines represent the approximate stress level at which deformation was measured. A linear regression is fit to the deformation versus stress data. The pyroplastic index is the amount of deformation a body exhibits at a stress level of 50 kPa. The deformation at 50 kPa is interpolated using the linear regression.

Proposed Standard Test Method

The proposed standard test method is shown in Table III.

Deformation Mechanisms

According to data presented here, a significant portion of deformation occurs before the dwell period at high temperature (typically 1250–1300°C). The onset of deformation occurs near 1150°C and continues at a much higher rate than deformation observed during the dwell period. The prominent deformation mechanism is viscous creep. Viscous creep in the low-temperature region occurs at a high rate, but at the peak temperature, creep occurs at a relatively slow rate. It is proposed that the low-temperature deformation event occurs during a non-steady-state condition. The

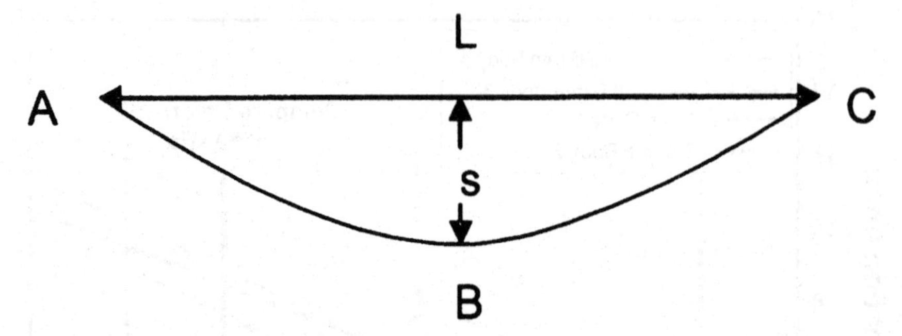

Figure 13. Variables in the parabolic arc length equation.

microstructure changes because of the increasing temperature, formation of mullite, and dissolution of quartz. At the peak temperature, it is proposed that a steady-state condition exists after a relatively short amount of time. According to recent literature, the glass phase composition at the peak temperature is constant after 1–2 h.[10–12] Therefore, after a dwell of 1–2 h, the deformation rate will remain constant because the porcelain microstructure remains constant. Steady-state creep is a result of deformation occurring through an unchanging microstructure at the peak temperature. Deformation data is used to calculate strain of the sample. The strain and previously calculated bending stress are used to calculate the system viscosity with the steady-state creep equation.

Strain

Span length and deformation are used to calculate the arc length (Fig. 13) of the deformed body using a parabolic arc length equation:[13]

$$\text{arc length } ABC = \frac{1}{2}\sqrt{L^2 + 16s^2} + \frac{L^2}{8s}\ln\left(\frac{4s + \sqrt{L^2 + 16s^2}}{L}\right) \tag{12}$$

where $ABC \equiv$ arc length, $L \equiv$ span length, and $s \equiv$ measured deformation.

Strain is then calculated using

$$\varepsilon = \frac{\Delta L}{L} = \frac{ABC - L}{L} \tag{13}$$

where L is the span length.

Strain rate is calculated by taking time into consideration:

$$\dot{\varepsilon} = \frac{\partial \varepsilon}{\partial t} \tag{14}$$

Viscosity

Viscosity is calculated using the steady-state creep equation for viscous flow:[14]

$$\sigma = 3\eta_s \dot{\varepsilon} \tag{15}$$

where $\sigma \equiv$ stress (Pa), $\sigma_s \equiv$ system viscosity (Pa·s), and $\dot{\varepsilon} \equiv$ strain rate (s^{-1}).

This equation assumes that no microstructural changes are taking place. According to recent literature, after a minimum dwell time of 1 h, quartz dissolution ceases and mullite remains constant, therefore, the microstructure is not changing.[10–12] The stress value used for the steady-state creep equation is that defined previously through the theory of elasticity. Strain rate is calculated based on sample deformation as mentioned previously.

The data shown in Fig. 14 were collected using a measurement technique called digital time lapse photography (DTLP) developed by Noirot.[15] DTLP allows dynamic noninvasive measurement of the deformation process. The setup consists of a camera placed on a tripod in front of a tube furnace. An alumina rod provids a constant reference position during firing, therefore the distance between the deforming sample and the reference allows for deformation measurement. Notice that a significant amount of deformation occurs prior to the peak temperature. Approximately two-thirds of the total deformation occurs before the soak.

The deformation data were used for calculating system viscosity values with the steady-state creep equation. Deformation had some scatter, which was detrimental when calculating strain values. In Fig. 14, an overall trend of deformation is apparent; however, instantaneous slopes are variable. For example, some instantaneous strain data points had negative slopes, which is not in agreement with the overall trend line. A nine-cycle moving average was used to manipulate the deformation data for calculating viscosity.[5]

Calculated system viscosity using the method of the moving average of the strain (nine cycles) is illustrated in Fig. 15. It is important to note that the viscosity values calculated are system viscosity values, not just the glass phase viscosity values. In the low-temperature region (below 990°C) calculated viscosity values are artificial because there is no viscous phase in

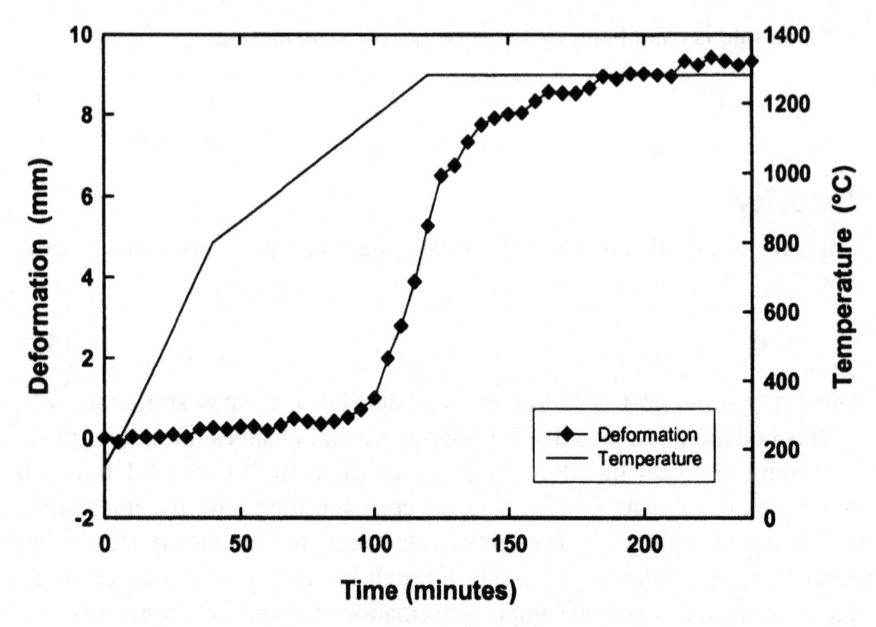

Figure 14. Deformation measured using the DTLP technique.

the system. The viscosity values begin to drop dramatically at approximately 990°C and at approximately 1190°C the rate of decreasing viscosity lowers. The minimum in viscosity is reached at approximately 1200°C. The region in which viscosity is low is in the same time frame as when a significant amount of deformation occurs.

Figure 15 also shows the dependence of stress on densification of the samples. Viscosity was calculated and plotted using both instantaneous stress (instantaneous density and dimensions) and constant stress (using fired density and dimensions). The viscosity is not significantly affected by using constant versus instantaneous stress values.

Conclusions

The stress on a body can be calculated using fired density, geometry, sample dimensions, span length, and force due to gravity. The stress versus deformation trend is used to establish a pyroplastic index of a body. The pyroplastic index is defined as the deformation at a stress of 50 kPa and successfully demonstrates the tendency of a body to deform.

According to deformation data collected during sample firing, two defor-

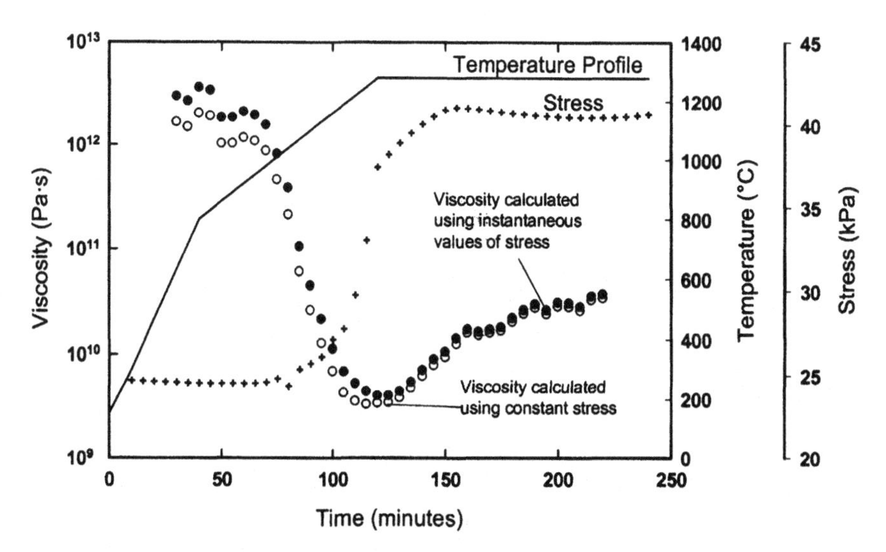

Figure 15. Calculated system viscosity using the steady-state creep equation.

mation events take place. The mechanisms are both due to viscous flow, but occur at much different rates. A high rate of deformation is apparent at low firing temperatures, proposed to be due to the microstructural evolution of the system. This deformation occurs in a non-steady-state condition. A secondary deformation event known as creep occurs at the dwell temperature. According to recent data, quartz dissolution is complete after 1–2 h at the peak dwell temperature, and mullite level remains constant. Therefore, it is proposed that the composition reaches steady-state. The steady-state creep equation is used to calculate the system viscosity.

References

1. I. C. McDowall and W. Vose, "Determination of Pyroplastic Deformation in Firing of Ceramic Bodies," *Br. Ceram. Trans.*, **50** [11] 506–516 (1951).
2. D. Seymour, "Phase Evolution in Electrical Porcelains During Firing," M.S. Thesis, Alfred University, Alfred, New York, 2000.
3. R. J. Roark, *Formulas for Stress and Strain,* 3rd ed. McGraw-Hill, New York, 1954. P. 92.
4. J. P. D. Hartog, *Strength of Materials.* Dover, New York, 1961.
5. A. Buchtel, "Pyroplastic Deformation of Whitewares," M.S. Thesis, Alfred University, Alfred, New York, 2003.
6. S. Timoshenko, *Theory of Elasticity,* 1st ed. McGraw-Hill, New York, 1934.
7. R. C. Hibbeler, *Mechanics of Materials,* 3rd ed. Prentice Hall, Upper Saddle River, New Jersey, 1997. P. 292.

8. R. J. Roark, *Formulas for Stress and Strain,* 3rd ed. Prentice Hall, McGraw-Hill, New York, 1954. P. 102
9. "Standard Test Methods for Apparent Porosity, Liquid Absorption, Apparent Specific Gravity, and Bulk Density of Refractory Shapes by Vacuum Pressure," ASTM Designation C830-00. American Society for Testing and Materials, West Conshohocken, Pennsylvania.
10. W. M. Carty, "Observations of the Glass Phase Composition in Porcelains," *Ceram. Eng. Sci. Proc.,* **23** [2] 79–94 (2002).
11. B. Earley, "Effect of Time on Quartz Dissolution in Porcelain Bodies," B.S. Thesis, Alfred University, Alfred, New York, 2002.
12. B. Pinto, "The Effect of Quartz Particle Size on Porcelain Strength," M.S. Thesis, Alfred University, Alfred, New York, 2001.
13. M. R. Spiegel, *Mathematical Handbook of Formulas and Tables.* McGraw-Hill, New York, 1968. P. 7.
14. W. D. Kingery, H. K. Bowen, and D. R. Uhlmann, *Introduction to Ceramics.* John Wiley & Sons, New York, 1976. P. 755.
15. M. Noirot, "Dynamic Pyroplastic Deformation Study: Digital Time Lapse Photography of Porcelain Firing," *Ceram. Eng. Sci. Proc.,* (2003).

Novel Casting Techniques for Whitewares

Philip R. Jackson
Ceram Research Ltd.

Casting using plaster molds is a long-established technique for producing complex-shaped items in the traditional ceramics sector. However, this process requires careful control and suffers from number of drawbacks,[1–3] including:

- Short lifetime of molds.
- Energy associated with drying molds during preparation and between casts.
- Labor and factory operating costs associated with retaining a plaster mold workshop.
- Faults in cast ware that are at least partly attributable to plaster molds (e.g., variable cast wall thickness due to variable moisture content in the mold; pinholing; poor cast surface from older molds that have roughened due to partial $CaSO_4 \cdot 2H_2O$ dissolution, etc.).

Although pressure casting using porous plastic molds has proved to be a viable alternative for large producers, there is still believed to be a niche for a novel casting technique capable of offering both flexibility in production and a low capital equipment cost option for the small to medium enterprise.*

Ceram Research has been awarded a Sustainable Technology Initiative (STI) project† to investigate the viability of casting ceramics by importing the technology of rotational molding from the polymer industry. A main advantage of this technology lies in the use of nonporous molds. This project will run between May 2002 and May 2005, ending with pilot plant proving trials at Ceram Research. Ceram Research is working with two

*In the European Union, the small-to-medium enterprise (SME) is defined as a company with up to 250 employees, an annual turnover of less than 40 million euros, and no greater than 25% ownership by a large (non-SME) company.

†STI is a scheme funded by the Department of Trade and Industry and the Engineering and Physical Sciences Research Council in the UK. STI provides funding for projects aiming to generate new products and processes that have a positive impact on the environment while also delivering step changes in economic performance (productivity, throughput, etc.). Consortia must involve universities, research centers, and industrial partners.

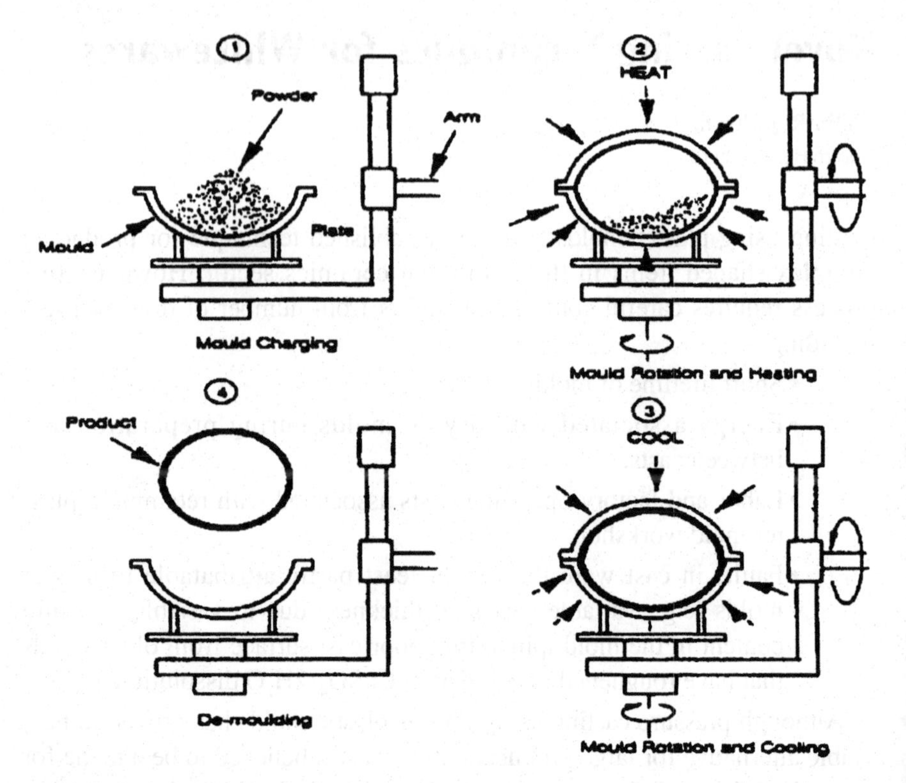

Figure 1. An illustration of the stages associated with rotational molding of plastics (sequence runs clockwise from top left).

other science providers (Loughborough University and Queens University, Belfast) as well as five whitewares producers, two material suppliers, and an engineering company.

The principles of rotational molding can be understood by referring to Fig. 1, which illustrates molding of a hollow egg-shaped item from a granular polyethylene feedstock. The lower part of a two-part mold is filled with a fixed weight of the granulate polymer and the upper part of the mold is then fixed in place. The closed mold is rotated (at very low rates; typically 0–8 rpm) on two axes while heating. As the temperature rises, the polymer starts to melt and adheres to the mold and/or other particles to provide an even coating of the inner mold walls. Once the polymer particles have fully melted to create a homogeneous, void-free casting, the rotating mold is cooled prior to demolding.

The concept with ceramics is to employ an extremely high solids content slurry instead of a powder. Such slurries potentially open up a range of processing routes[4] in which solidification is triggered after pouring into a mold. However, investigations at Ceram have focused on using temperature-induced chemical changes[5–7] that cause solidification by coagulation.

Attaining a high solids content slurry depends first on controlling the zeta potential of the components in suspension. Zeta potential recognizes that a powder particle in water often assumes a positive or negative surface charge due to ion dissolution/adsorption at the particle surface. The larger this charge, the greater the size of the surrounding "double layer"[8] containing ions of the opposite charge. Where the surface charge is deemed inappropriate, surfactants can be used to alter the surface. A large double layer encourages particles to keep away from each other and so generates a lower viscosity slurry. However, at the very high solids levels required for this project, the particles are very close to each other and the use of a purely electrostatic surfactant to raise zeta potential can crowd particles together and actually raise viscosity. A better approach is to employ electrosteric surfactants where both surface charge and polymeric steric effects help to keep particles apart.

Multicomponent slurries (such as the clay/mineral flux/quartz mixtures employed in tableware and sanitaryware) offer a big challenge in that the different components have a different zeta potential when dispersed in water (see Fig. 2). Recognizing these differences and altering individual zeta potentials prior to mixing can help maximize final solids loadings in pourable mixtures.

The second factor that is critical in maximizing solids content in ceramic slurries is particle size distribution. German[9] reviews some of the approaches taken by different investigators (such as Grafton and Fraser, Furnas, Andreason) to address the issue of optimizing particle packing. Starting with theories based on perfect spheres, the text goes on to describe the influence of factors such as particle shape, surface roughness, and agglomeration. Funk and Dinger[10] have developed an equation for defining an ideal continuous particle size distribution based on defined minimum and maximum particle sizes (D_S^n and D_L^n, respectively, in Fig. 3). If we apply the equation to a typical tableware body, it can be appreciated how body slurries used in conventional plaster casting deviate significantly from "ideal" packing. Many theories on particle packing of course are based on dry spheres; investigators such as Ortega[11] have extended such principles into powder suspension scenarios by recognizing the concept of interparticle

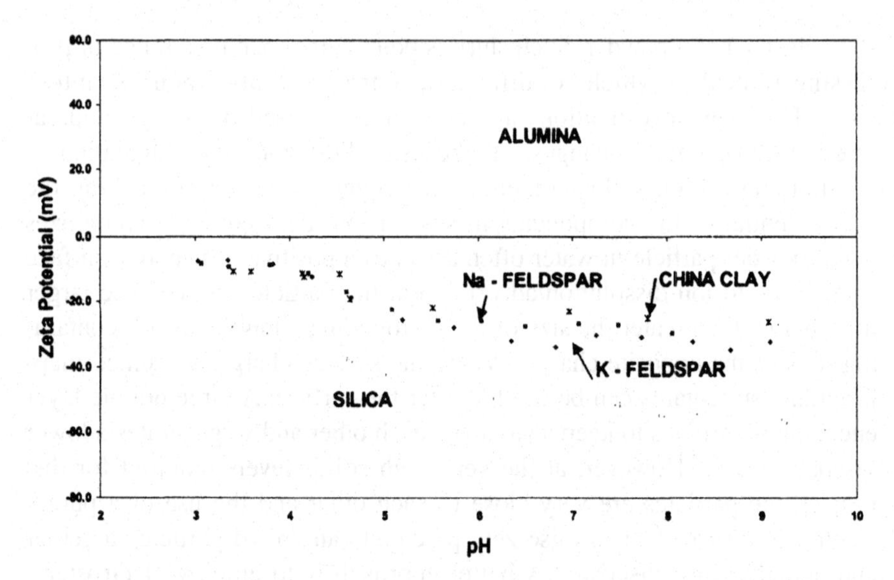

Figure 2. Zeta potential plotted against pH for various mineral phases found in whitewares suspensions.

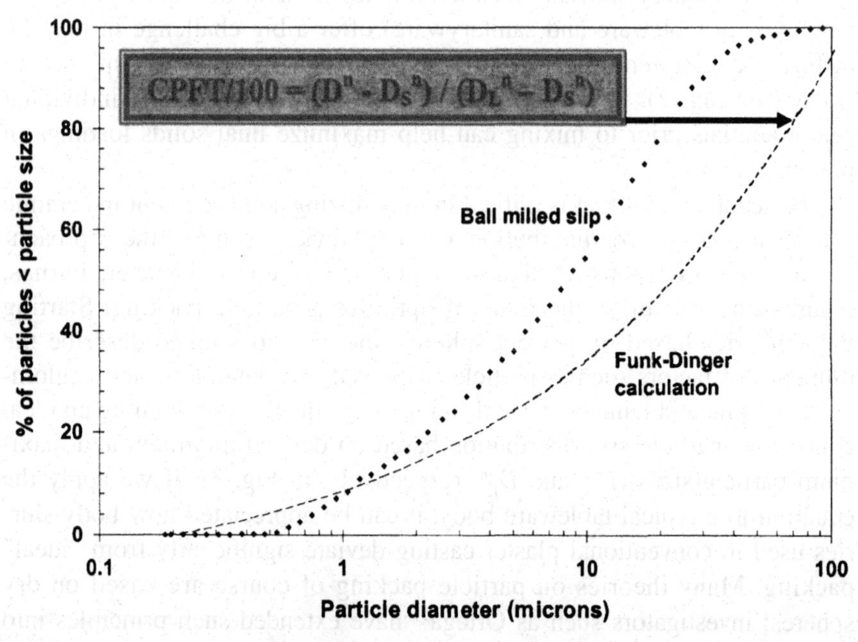

Figure 3. Actual and ideal (calculated) particle size distributions for a whiteware body.

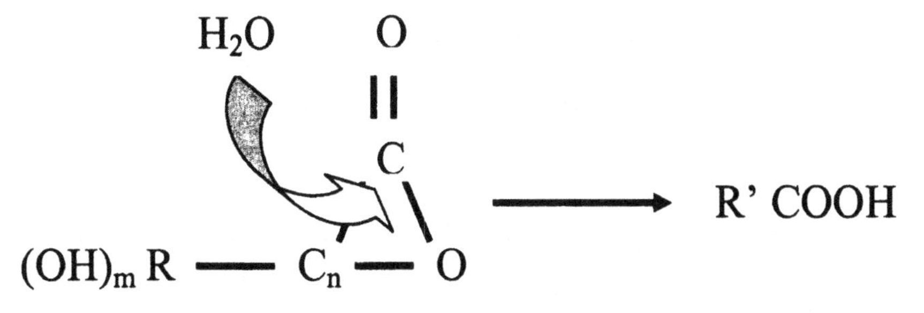

Figure 4. Lactone hydrolysis to create H^+ for coagulation casting.

spacing (IPS) in which water layers around particles control particle interaction and thus viscosity.

Once the solids content for a given body slurry has been maximized, attention can be turned to the methodology behind coagulating or solidifying the slurry once it is coating the insides of the molds. Ceram Research have been investigating two options,[5–7] although both involve the use of temperature-induced chemical reactions to create H^+ and so reduce the negative zeta potential present in the starting slurry.

The first chemistry being employed is that proposed by Gauckler et al.[5]

The second chemistry is based on lactones and is patented by Dytech.[6,7] At elevated temperatures, hydrolysis of the lactone (see Fig. 4) releases acid, which will lower the zeta potential (drag it toward the isoelectric point) of negatively charged particles in suspension. The project is essentially split into three phases:

- Phase 1 (May 2002–September 2003): Suspension optimization.
- Phase 2 (September 2003–August 2004): Identifying rotational molding conditions and definition of rotational molding equipment design.
- Phase 3 (August 2004–May 2005): Pilot trials.

The status of the work can be summarized thus:

- A pourable four-component body slurry with a composition relevant to the vitreous hotelware and sanitaryware industries has been prepared at a solids loading of >2 $kgdm^{-3}$.
- Solid casting (i.e., without rotational molding) has been achieved with suitable lactones. Early results suggest that the green strength of the coagulation cast bars are comparable to conventional plaster cast bars.

- Selection of lactone chemistry, concentration, and temperature have a marked effect on setting times and require careful optimization.
- Rotational molding experiments are due to begin during October 2003 in Belfast. Feedback from these experiments will lead to the design of an experimental rig capable of casting between one and six small items.

References

1. M. E. Twentyman and P. Hancock, "The Causes of Pinholes in Vitreous Sanitaryware." BCRA research paper No. 706, 1979.
2. I. Berger, I. Seidermann, and A. Baumgarten, "The Formation of Casting Pores in Slip Cast Products: Means of Prevention," *CFI/Berichte DKG,* **67** [6] 239 (1990).
3. D. Basnett, "The Control of Casting Slip: A Survey of Current Casting Practice in the Tableware Industry." Ceram research paper 749, 1987.
4. W. M. Sigmund, "Novel Powder-Processing Methods for Advanced Ceramics," J. Am. Ceram. Soc., **83** [7] 1557–1574 (2000).
5. F. H. Baader, T. J. Graule and L. J. Gauckler, "Direct Coagulation Casting — A New Green Shape Technique, Part II: Application to Alumina," *Indust. Ceram.,* **16** [1] (1996).
6. R. M. Sambrook, J. G. P. Binner, J. Davies, and A. J. McDermott, "Coagulation to make dense ceramics." International patent WO 96/04219, 1996.
7. R. M. Sambrook, J. G. P. Binner, J. Davies, and A. J. McDermott, "Making Dense Ceramics by Coagulation." European patent EP 0773913, 2001.
8. "Molecular Aspects of Clay-Water Interactions"; in *Clay-Water Interface and Its Rheological Implications.* Edited by N. Guven and R. M. Pollastro. Clay Minerals Society Boulder, Colorado, 1992.
9. R. M. German, *Particle Packing Characteristics.* Metal Powder Industries Federation, 1989.
10. D. R. Dinger and J. E. Funk, "Particle Packing III: Discrete versus Continuous Particle Sizes," *Interceram,* **41** [5] 332–334 (1992).
11. F. S. Ortega, R. G. Pileggi, A. R. Studart, and V. C. Pandolfelli, "IPS: A Viscosity-Predictive Parameter," *Bull. Am. Ceram. Soc.,* **81** [1] (2002).

Effects of Borax Solid Wastes Fritted and Added into Wall Tile Opaque Glazes on the Final Mirostructure

G. Kaya
Anadolu University, Turkey

The boric acid content of a commercial opaque frit used in the preparation of wall tile glazes was replaced by borax solid wastes from Etibor Kirka Borax Company of Turkey in order to decrease costs. In addition to the effects of such modification on the service performance of the relevant glazes, changes in the final microstructure were also investigated with the aid of certain characterization techniques. This study confirmed the potential usage capacity of these wares in the form of frit and evaluated their use in the production of wall tile opaque glazes.

Quartz Dissolution into Porcelain Glasses

Caspar J. McConville, Amit Shah, and William M. Carty
Whiteware Research Center, New York State College of Ceramics at Alfred University, Alfred, New York

Introduction

Porcelains consist of crystalline material (quartz, mullite, alumina) surrounded by large quantities (≈60%) of amorphous phase. There are different opinions in the literature as to the composition of the amorphous material in porcelains: some authors think it has a variable composition, and others that it is homogeneous in a given sample. On firing, quartz particles are dissolved into the glass phase, leaving amorphous "solution rims" in the microstructure. These glassy regions are found surrounding the remains of large quartz grains, which have almost completely dissolved, and tend to separate the quartz from the surrounding mullite and alumina crystals. These areas of the microstructure allow chemical analysis of discrete amorphous areas, unimpeded by crystals that are otherwise predominant in the amorphous phase.

For this research investigation, a model system was set up in the form of a diffusion couple, an experiment in which a silica-deficient glass is placed in contact with a quartz crystal and heated. During heating, the glass tends to dissolve the quartz and should theoretically result in a silica-rich amorphous phase at the glass-quartz boundary. In addition, actual solution rims in an industrial porcelain system were subjected to microstructural examination to determine their compositions.

Materials and Methods

A diffusion couple experiment was performed to analyze the dissolution of quartz by glass phases on firing. A glass with a low silica level was made (chemical analysis shown in Table I), and one face of it was polished. A quartz crystal was sliced, and one surface of it was polished and then scored with three channels. Platinum wires were placed into the channels in the quartz. These wires marked the original position of the boundary between glass and quartz. The experimental setup is shown in Fig. 1. The polished faces of the glass and quartz were placed together, and the specimen was fired for 3 h at 1200°C in an electric furnace. On cooling, the glass boundary had advanced past the platinum wires into the quartz. A cross section of the diffusion couple was prepared for scanning electron microscope (SEM) and energy dispersive X-ray spectrometry (EDS) analysis by sectioning and polishing. The SEM used was a Philips 512 fitted with an Evex quantitative EDS system.

Table 1. Chemical composition of the experimental glass

Oxide	Wt%
SiO_2	65.5
Al_2O_3	19.78
K_2O	11.14
Na_2O	2.79
CaO	11.14
Fe_2O_3	0.1
TiO_2	0.05
MgO	Trace
P_2O_5	Trace

For quartz solution rim studies, industrial porcelain materials were used that had been manufactured using standard industrial techniques. Two materials were used for the study: a high-alumina whiteware body and a high-silica insulator body porcelain. Samples were prepared for transmission electron microscope (TEM) analysis by cutting a thin section (≈500 μm) using a microtome, and mechanically grinding and polishing to ≈30 μm before argon-ion milling to electron transparency, and carbon coating. TEM analysis was carried out at an accelerating voltage of 120 kV in a Jeol 2000-FX TEM, with a PGT EDS system.

Results and Discussion:

Figure 2 shows a secondary electron SEM image of the prepared diffusion couple specimen. The original silica-deficient glass can be seen to the left, the platinum wire marking the original boundary labeled, and the original and new boundaries are marked.

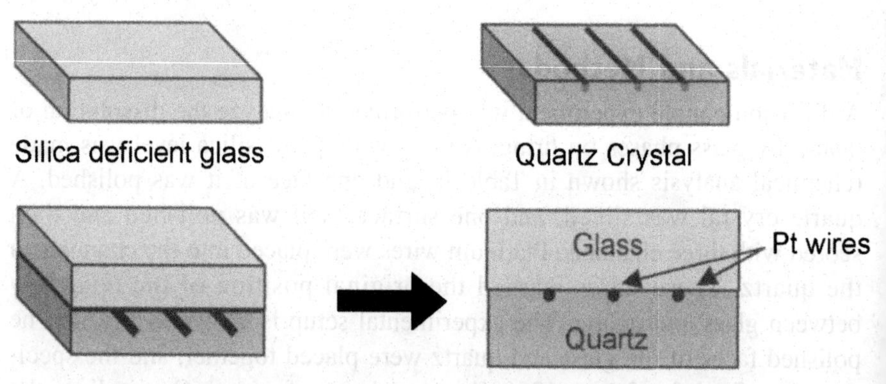

Figure 1. Experimental setup for the glass/quartz diffusion couple experiment.

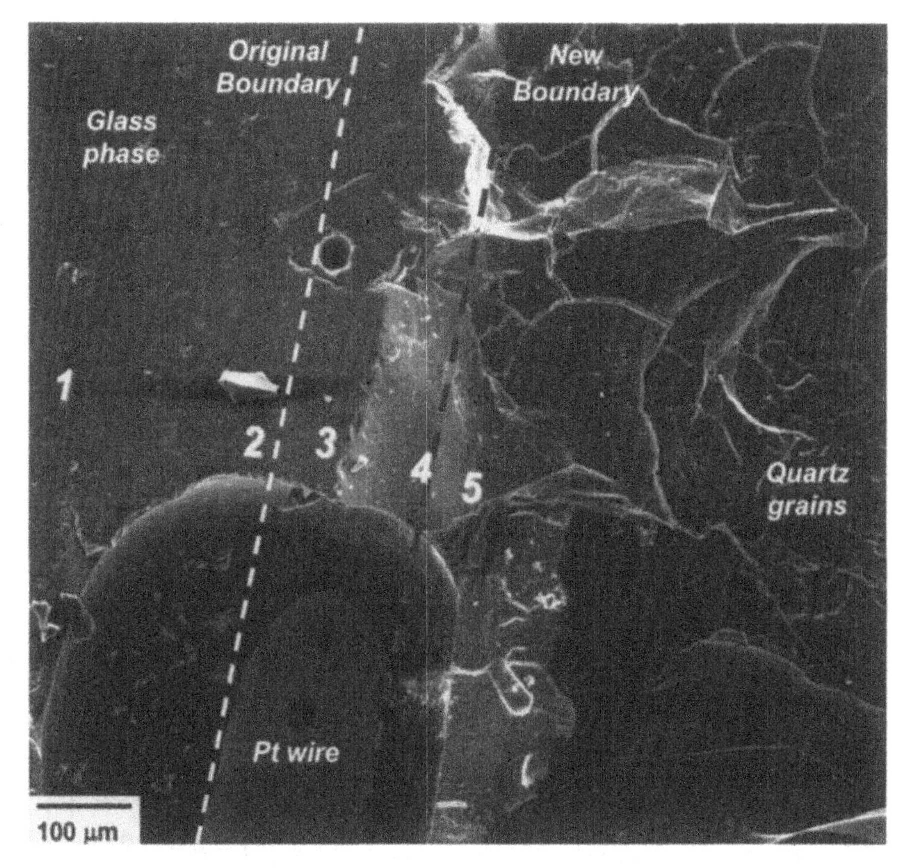

Figure 2. SEM micrograph of fired diffusion couple specimen.

The quartz crystal can be seen to the right. The glass boundary has moved more than 100 µm into the quartz crystal during heating. Quantitative EDS analysis was carried out at the points labeled 1–5 in Fig. 2. Table II summarizes the data from the EDS analysis, in addition to a calculated composition based on the original batch recipe for the glass. The data in the table are also presented graphically in Fig. 3. The percentage of silica in the glass increases from Point 1, which is close to the theoretical glass composition and 200 µm from the original quartz boundary, to Point 5, which has nearly 20% more silica and is beyond the quartz boundary. The K_2O and alumina levels also fall across the newly created glass region. However, the proportion of Na_2O increases steadily as the new quartz boundary is approached, indicating migration of this flux toward the boundary.

Table II. Glass phase composition calculated from batch recipe, and measured from points 1 to 5 in Fig. 2 by SEM/EDS

Oxides	Calculated	Point 1	Point 2	Point 3	Point 4	Point 5
SiO_2	65.5	63.71	62.65	64.51	81.51	82.41
Al_2O_3	19.78	20.21	19.3	17.98	7.7	7.43
K_2O	11.14	15.26	16.5	15.07	8.06	6.76
Na_2O	2.79	1.36	1.55	2.44	2.74	3.4

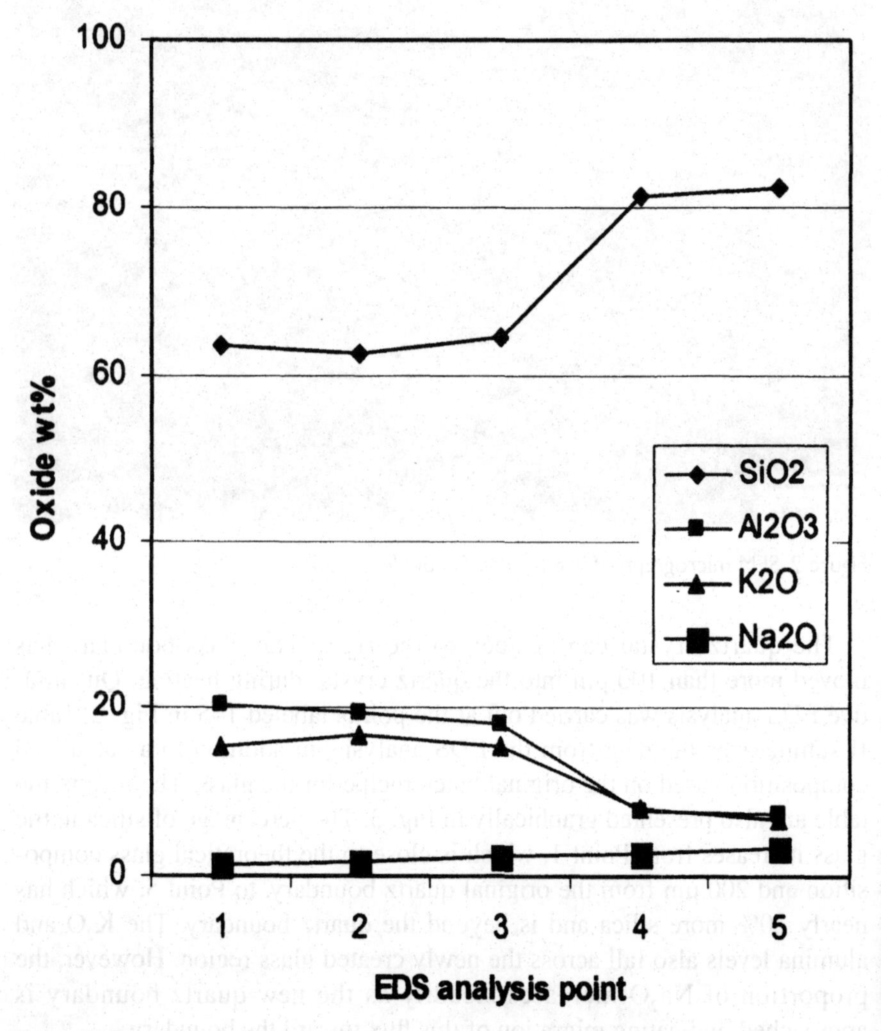

Figure 3. Chemical composition of the glass at points 1–5 in Fig. 2.

The TEM study involved two porcelains, a high-alumina whiteware and a high-silica insulator body. These were chosen to give the potential for variation in the glass phase produced. The study concentrated on EDS analysis of the solution rims around quartz grains in the fired bodies. Figure 4 shows the TEM and EDS results for the high-alumina body. The brightfield TEM image shows a relict quartz grain with bend contours, surrounded by an amorphous region that is free of crystals. This is the "solution rim" that results when the original large quartz particle is dissolved by the surrounding glass. Previous research has suggested that the composition of this region is pure silica. EDS spectra were taken at the points indicated on the photograph. The results show that even at Point 2, which is only 50–100 nm from the quartz grain and 1 μm from the nearest alumina crystal, some alumina and potassium are present in the glass. This indicates flux migration toward the quartz boundary and suggests that the formation of the solution rim is dependent on the presence of a flux in this region. Figure 5 shows a similar solution rim in the high-silica body. Again, alumina, sodium, and potassium are found throughout the glass phase, up to the quartz boundary.

Summary

The diffusion couple experiment showed that a low-silica glass will attack and erode a quartz particle when fired to 1250°C.

Compositional analysis indicated that alumina and alkalis were present in all the glass formed, even that close to the boundary. This indicates that alumina has to diffuse through the amorphous phase along with the alkalis to allow a glass to be formed. A linescan showed silica levels in the glass increasing slightly 30 μm from the boundary in the diffusion couple.

TEM images showed different feldspar-relict mullite morphologies in different porcelains.

TEM/EDS analysis of the quartz solution rims indicates an almost uniform composition between the quartz-glass interface and the mullite-glass interface.

Alumnia is seen in all the glass formed, even at the quartz-glass interface.

This argues in favor of a uniform glass phase composition in a given porcelain at a given firing temperature.

Acknowledgment

SEM/EDS for the diffusion couple specimen was performed by Ward Votava.

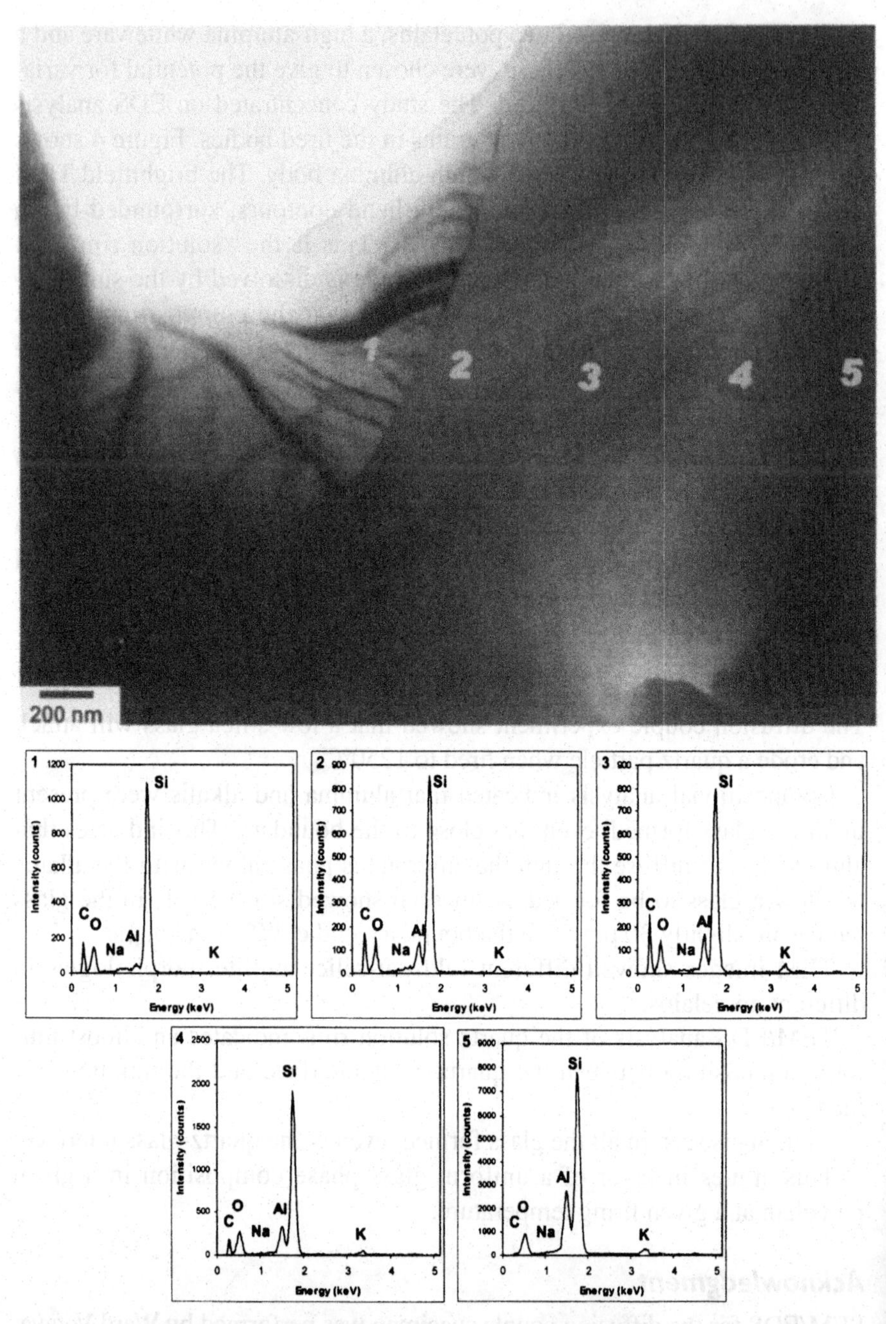

Figure 4. Brightfield TEM image and EDS analyses of a solution rim in the high-silica insulator body, showing a quartz crystal surrounded by amorphous material. Points 1–5 indicate where EDS spectra were taken.

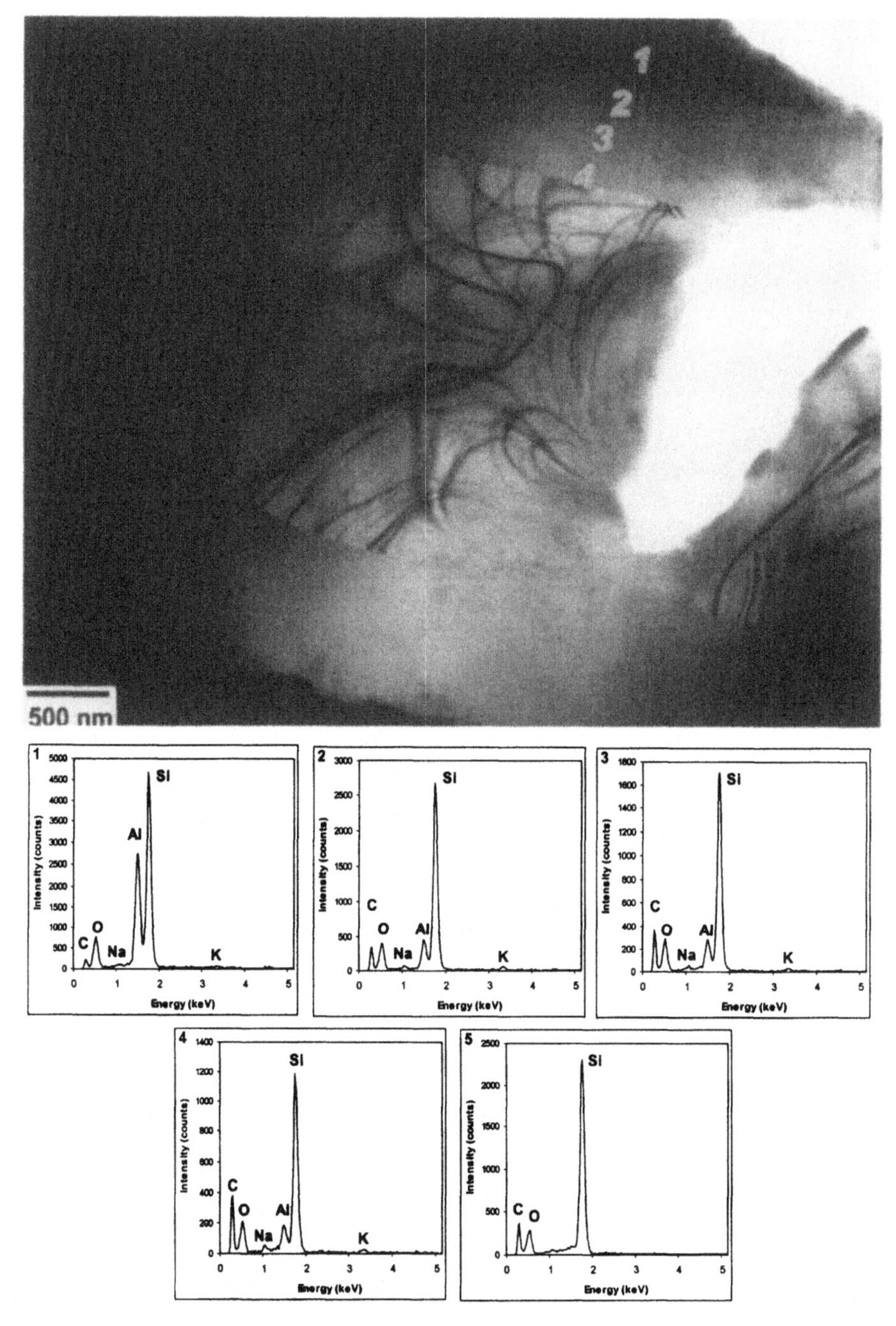

Figure 5. Brightfield TEM image and EDS analyses of a solution rim in the high-alumina porcelain whiteware body. EDS spectra were taken from points 1–5 marked on the image.

Role of Polymeric Additive Compatibility in Ceramic Processing Systems

U. Kim and W. M. Carty
New York State College of Ceramics at Alfred University, Alfred, New York

The interactions between polymeric additives in ceramic processing are usually considered negligible. In this study it is shown that PVA binder migration during the spray drying process is influenced by the choice of dispersant. This result suggests that interactions between polymers usually do occur and can adversely affect product performance. Flory-Huggins calculations predict phase separation of Na-PMAA and Na-PAA with PVA and homogeneous mixing of Na-silicate with PVA. Light scattering studies on polymer solutions and morphology studies on dried polymer solutions confirm these predictions. Based on these calculations and the observations, PVA binder migration behavior is explained. PVA in spray-dried granules can be stained and its location within the spray-dried granule observed, verifying the effect of other polymers (dispersants) on binder migration.

Effect of Die Fill on Compaction of Granular Bodies

Brett M. Schulz and William M. Carty
Whiteware Research Center, New York State College of Ceramics at Alfred University, Alfred, New York

Nikalos J. Ninos
Buffalo China Inc., Buffalo, New York

Three binder systems, which exhibited a range of P1 values, were used to investigate the effects of die fill on the compaction of granulate. The binders selected for this study were a plasticized PVA, sodium lignosulfonate, and PEG 8000. For comparison, granulate was prepared without an organic binder system. Samples were pressed in an industrial semi-isostatic horizontal dry press and evaluated for surface roughness using an optical interferometer. A comparison to samples prepared using the plasticized PVA binder system in a vertical dry press indicated that die fill has a major role in good compaction. The samples prepared using the horizontal dry press had a consistently higher surface roughness due to poor die fill prior to compaction.

Introduction

Organic binder systems have been used extensively in the manufacture of ceramic ware to improve green strength. Spray drying of ceramic powders is used to further improve productivity by increasing the flowability of the powder for good die fill prior to compaction. In this study the surface roughness of dry-pressed ware manufactured on a horizontal press (a vacuum system is used to fill the die prior to compaction) is compared to the ware from a vertical press (a gravity-assisted vacuum system is used for die fill). Three organic binder systems were evaluated on the horizontal press to determine the effect of binder systems that exhibit a range of pressure at the onset of granule deformation (traditionally referred to as P1 in the compaction diagram).

To better understand the mechanics of granule deformation during compaction, the effects of moisture content and granule size have been revisited. As a comparison, the deformation of granulate prepared without an organic binder system is evaluated.

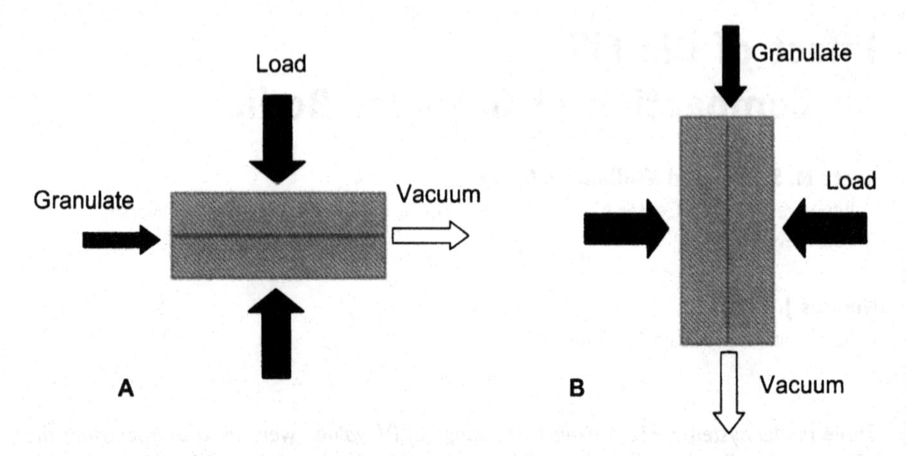

Figure 1. Schematic of the dies installed in the industrial presses used for this study, (A) horizontal press and (B) vertical press. The dies are shown as the cross-hatched boxes with the load applied as indicated by the gray arrows. The granulate is fed as illustrated by the black arrows and facilitated by a vacuum system (white arrow). In the case of the vertical press (B), the vacuum system is assisted by gravity to fill the die.

Background

The surface roughness of ware manufactured by dry pressing is significantly higher than that of ware produced by traditional plastic forming methods. Recent evidence has shown that phase separation of polymeric additives occurs during spray drying. The organic-rich film that is created on the surface of the granule leads to case hardening and poor deformation under pressure. The use of semi-isostatic presses to manufacture traditional ceramics (i.e., porcelain dinnerware) has gained significant interest because of the rapid pressing rates, the ability to form complex shapes, and the elimination of the slow drying stage inherent to plastic forming processes. The case-hardened granulate not only exhibits a high P1, but the presence of a compliant membrane, which distributes the load, further aggravates the problem. The complaint membrane deforms around the granulate, and relics remain in the pressed surface. Changes in the binder system used for preparing spray-dried granulate can significantly affect the measured P1 value, because of not only possible polymer phase separation but also the properties of the polymeric additives themselves. It was anticipated that changes in P1 could be used to predict variations in the measured surface roughness.

Table I. Organic binder systems used to compare surface roughness readings from the two industrial semi-isostatic presses

Designation	Binder system
A	Na-lignosulfonate
B	Celvol 203 PVA, Nalco 93QC215 plasticizer (standard binder system)
C	PEG 8000

Die Fill Effects

Procedure

Two industrial presses were used to manufacture dinnerware for evaluation: a horizontal semi-isostatic press, which uses a vacuum system to fill the die, and a vertical semi-isostatic press, which uses a gravity-assisted vacuum system to fill the die. The differences in these two presses are shown schematically in Fig. 1. Three organic binder systems have been evaluated on the horizontal semi-isostatic press; details of the binder systems are listed in Table I. Granulate B was also evaluated on the vertical press as a comparison. The total concentration of organic additives was maintained constant in each binder system. Settings on the horizontal press were unchanged for each binder system with the peak pressure of 280 bar (4060 psi). The operation of the vertical press was similar with the settings tailored to an industrial manufacturing process and the peak pressure at 300 bar (4350 psi).

The surface roughness of the pressed dinnerware was measured using an optical interferometer.* An explanation of the optical interferometer and the process used to collect and analyze the data is given elsewhere. Images consisting of 1.23 million data points were collected, corresponding to an area of 5.32 × 4.32 mm on the surface of the samples (a 5 × 5 grid of images). Fourteen samples were measured for each binder system prepared on the horizontal press and 20 samples were analyzed from the vertical press to develop statistically significant results. Results were compared using ANOVA to determine if a statistical difference existed in the results.

*NewView Model 5032, Zygo Corporation, Middlefield, Connecticut.

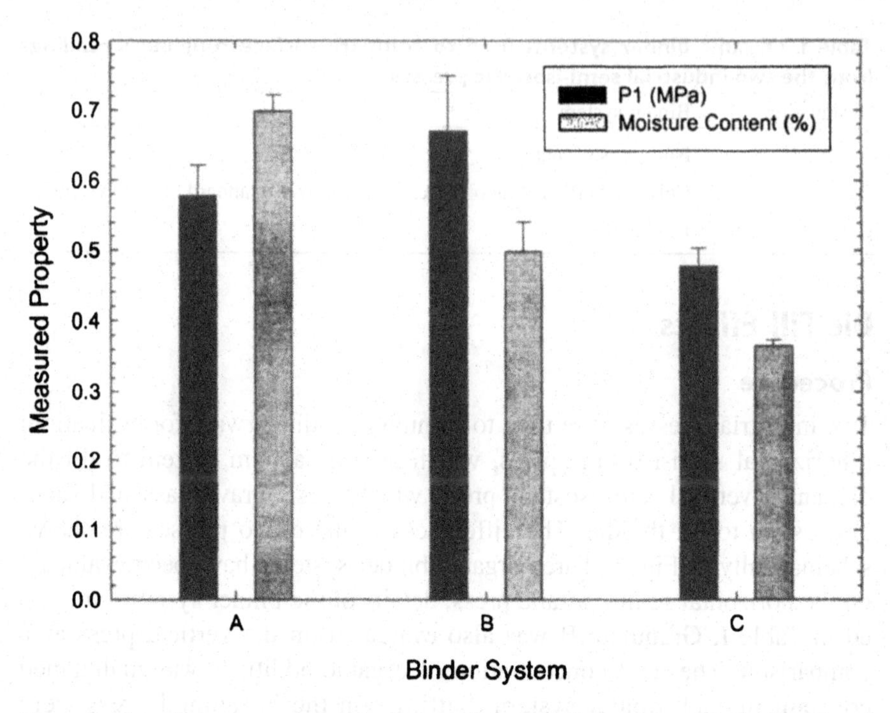

Figure 2. Plot of the pressure at the onset of granule deformation, P1, (in black) and the moisture content (gray) of the granulate that had been stored in a sealed container for 6 months prior to analysis.

Results

P1 for each binder system was determined using an Instron universal testing machine.[†] The results are shown in Fig. 2 for each binder system used in the die fill study. Two grams of granulate were loaded into a 0.75 in. inner diameter hardened steel die lubricated with oleic acid. The load was increased up to 8030 lb force at a crosshead speed of 0.11811 in./min (3 mm/min). It was predicted that softer binder systems — that is, systems with a lower P1 value — should deform better under semi-isostatic compaction.

Representative images, taken using the optical interferometer, from Granulate B (horizontal press) and Granulate B (vertical press) are shown in Fig. 3. The image on the left, from the horizontal press, indicates the presence of deep crevices in the sample surface, resulting in a RMS roughness of 11.7 μm. The image on the right, from the vertical press, shows that

[†]Instron Model 8562, Instron Corporation, Canton, Massachusetts.

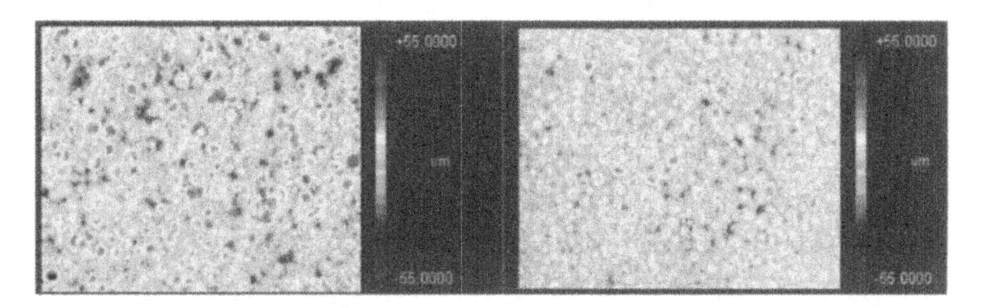

Figure 3. Representative contour plots of the dry-pressed surface from binder system B from the horizontal press (left) and the vertical press (right) taken using the optical interferometer. The RMS roughnesses were 11.7 μm and 8.2 μm, respectively.

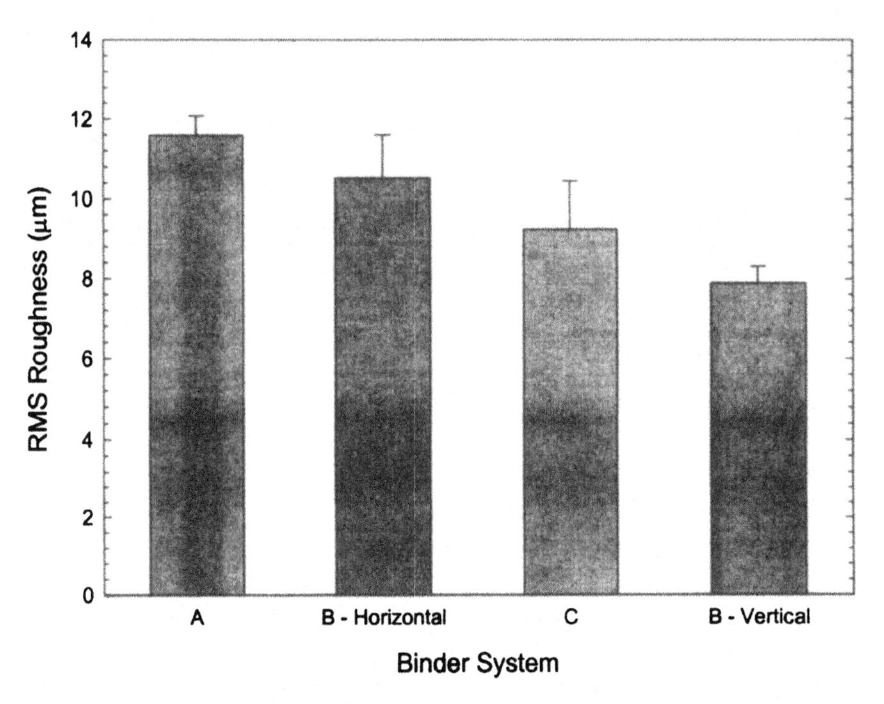

Figure 4. Plot of the RMS surface roughness from the three organic binder systems studied in the comparing the two industrial semi-isostatic presses.

while the surface still consists of peaks and valleys, it does not contain the deep crevices, reducing the RMS roughness to 8.2 μm. The average measured RMS roughness from the samples is shown in Fig. 4. The samples prepared on the horizontal press all have a significantly higher RMS surface

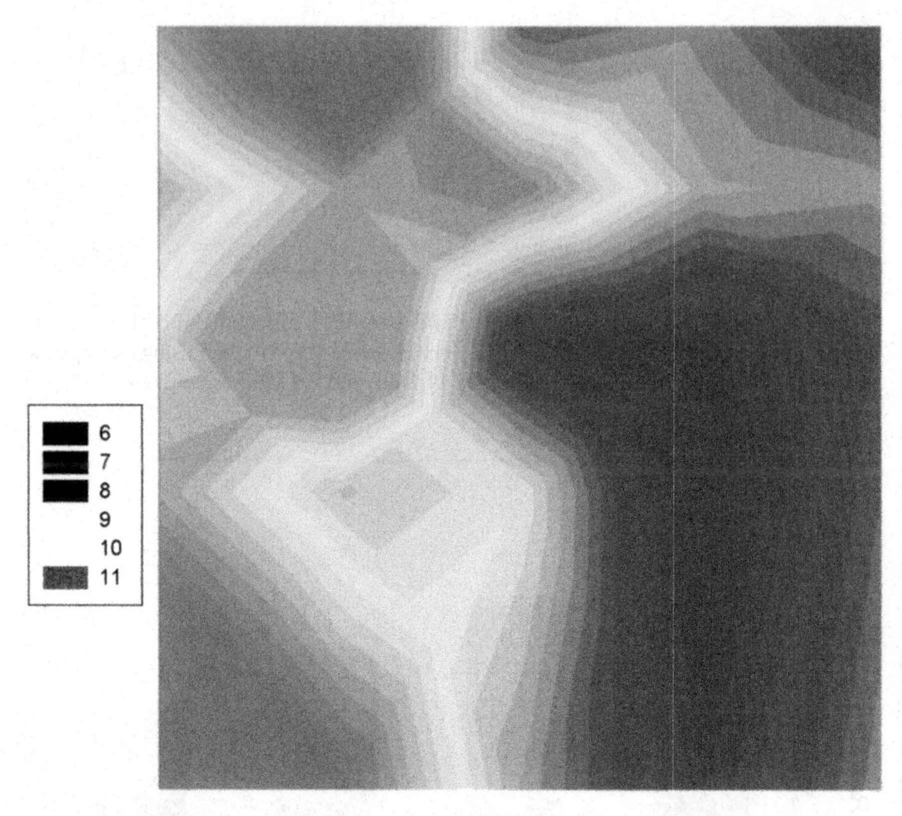

Figure 5. Contour plot of the RMS surface roughness as a function of position on the binder system B sample from the horizontal press.

roughness than the samples prepared using the vertical press. To determine the variation in the surface roughness, data from the interferometer were partitioned into the original 25 images. Each image was analyzed for RMS surface roughness. Contour plots of the RMS roughness as a function of position were created; see Fig. 5 (Granulate B, horizontal press) and Fig. 6 (Granulate B, vertical press). The variation in the RMS roughness from the horizontal press is significantly greater than that from the vertical press.

The roughness results do not follow the trend predicted from the measured P1 data: the ware manufactured on the vertical press has a significantly lower measured roughness despite the higher P1 of Granulate B. Die fill complications and poor die fill artificially inflate the measured roughness of the ware from the horizontal press because of the presence of large crevices

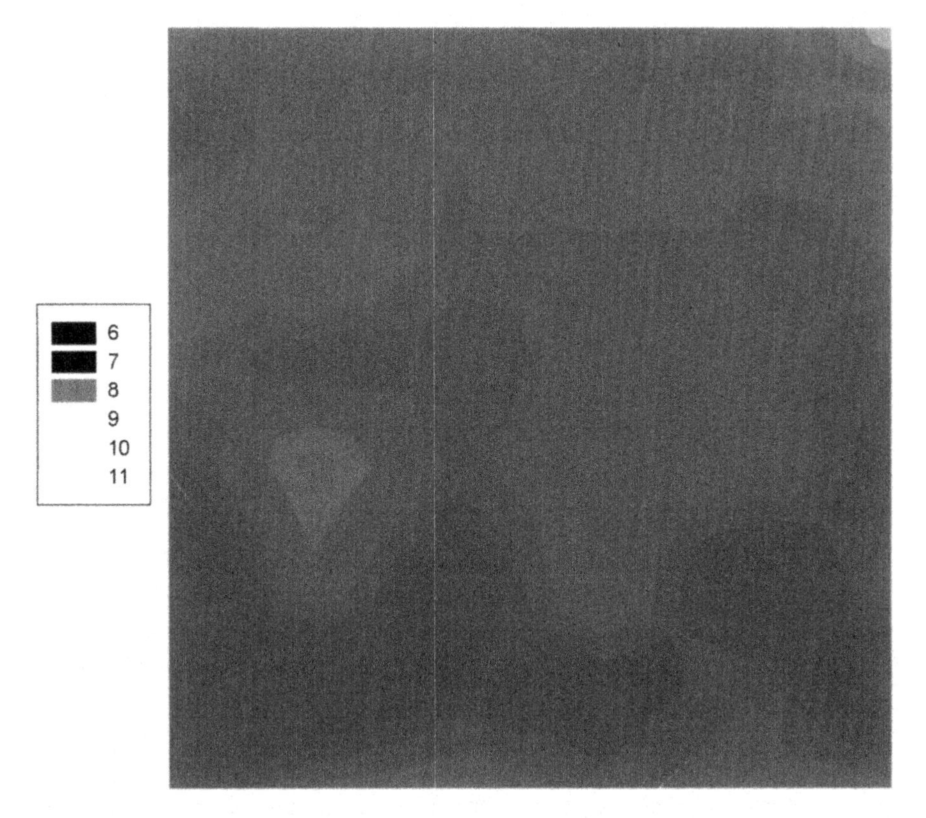

Figure 6. Contour plot of the RMS surface roughness as a function of position on the binder system B sample from the vertical press.

in the surface that were not removed during pressing. Therefore, the mechanics of granule deformation and the effect of moisture content and granule size were revisited.

Mechanics of Granule Deformation

Relative Humidity Effects

Procedure

In addition to the three binder systems already discussed, a fourth sample of granulate, with no added organic binder system, was evaluated to understand the mechanics of granulate deformation. This binder system is designated Granulate D.

Table II. Storage conditions used to regulate the relative humidity to rehydrate the granulate in this study

Storage condition	Storage temperature (°C)	Relative humidity (%)	Solubility (g/100 g H_2O)
As-dried	~18 (room temp.)	0	N/A
Humidity unregulated	50	5–10	N/A
LiCl	50	18	90
$NaNO_3$	50	60	95
KCl	50	90	85
H_2O	50	100	N/A
As-received	~18	N/A	N/A

Approximately 2 g of granulate was weighed into porcelain crucibles and the moisture content of each sample was determined by drying at 110°C for 24 h. The crucibles were then stored for 3 weeks at 50°C in desiccators above saturated salt solutions that controlled the relative humidity within the chamber. The higher the solubility of the salt on a mole fraction basis, the lower the relative humidity within the chamber. The various treatments for the granulate in this study are listed in Table II. The dried samples were not subjected to any treatment during the 3 weeks; the samples were stored in sealed plastic bags until they were tested. The relative humidity was not regulated during the 3 weeks of storage for the samples stored at 50°C and ambient humidity, designated as "humidity unregulated." After 3 weeks the samples were cooled to room temperature and the change in weight due to absorption of water was measured. The samples were sealed in plastic bags for an additional 2 weeks prior to determining the pressure at the onset of granule deformation; during this time the granulate reached an equilibrium with the surrounding atmosphere.

The onset of granule deformation for each binder system and humidity was measured by generating compaction diagrams on the Instron. Five samples were generated for each experimental condition in the study. After pressing the pellets in the Instron, the weight of the pellet was recorded and the samples were dried at 110°C for 24 h.

After drying, the thickness, diameter, and weight of the pellets were recorded to determine moisture content (at the time that the compaction diagrams were generated) and density of the pressed pellets. Green strength of the pellets was determined by diametric compression on the dried samples.

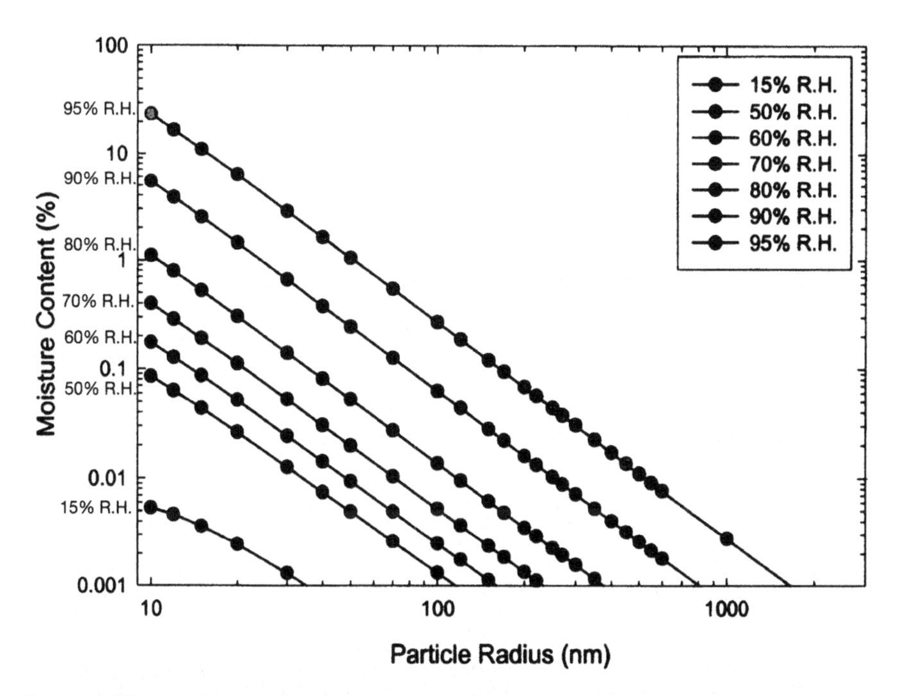

Figure 7. Plot of the calculated moisture content of a particulate bed as a function of the particle radius for several relative humidities. The calculation assumes FCC particle packing in the powder bed.

Results

Storing granulate in humid surroundings allows physical water to condense in the pore structure of the granulate, a process that is controlled entirely by thermodynamics. By making assumptions about the particle packing within a powder bed (or a dried granule), it is possible to calculate the moisture content as a function of particle radius. In the example shown in Fig. 7, a face-centered cubic packing structure of monosized particles was assumed and the amount of physical moisture condensed at the particle contacts was calculated. From this model it is predicted that a powder bed of 10 nm monosized particles will become saturated with physical water at 95% relative humidity.

A plot of the cylindrical pore radius filled as a function of the relative humidity is shown in Fig. 8. Also shown in Fig. 8 is the moisture content of the granulate after 3 weeks of storage at 50°C over the salt solutions used in this study. A similar trend is seen in the data indicating that the neck region

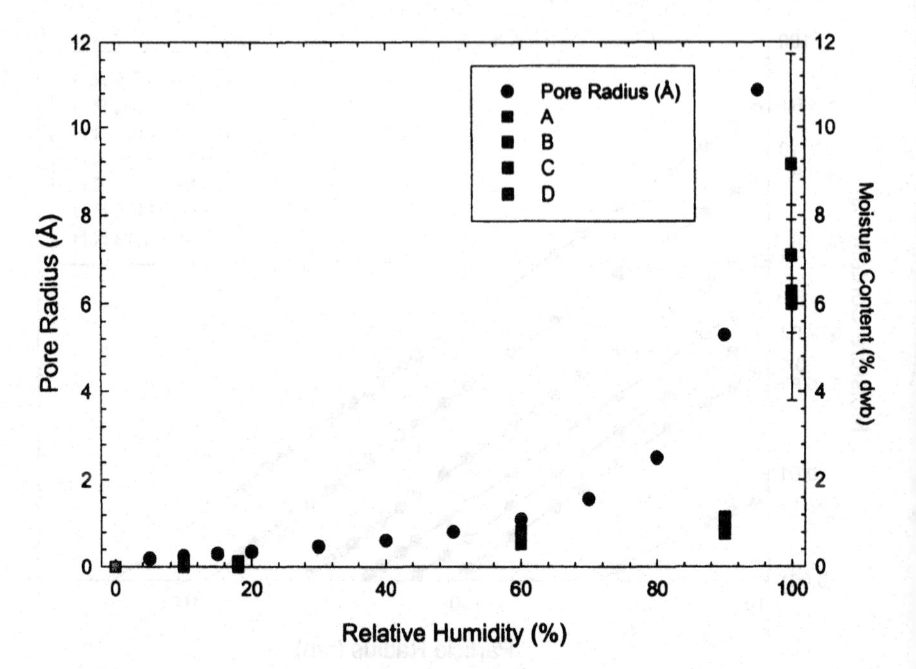

Figure 8. Plot of the pore radius filled as a function of relative humidity of the surrounding environment and the moisture content of the granule samples after 3 weeks in a humidity-controlled environment.

between particles in the granulate has become filled with physical water. As the relative humidity increases, more physical water is condensed in the structure; deviations between the two curves are artifacts of noncylindrical pores in the spray-dried granulate. During the 2 weeks that the hydrated granulate was stored at ambient conditions prior to determining the onset of granule deformation, there is a change in the moisture content as the granulate reaches equilibrium with the surrounding ambient conditions. The moisture content of the granulate after generating the compaction diagrams is plotted in each figure.

The results for the binder systems used in this study are summarized in Fig. 9 (P1), Fig. 10 (pressed density), and Fig. 11 (green strength) as a function of the relative humidity at which the granule was stored. The P1 value is significantly different in the range between 10 and 60% relative humidity. In this range the P1 value for Granulate B is significantly lower than that of Granulate A. Little difference is seen in the pressed density and the green strength between the granulate.

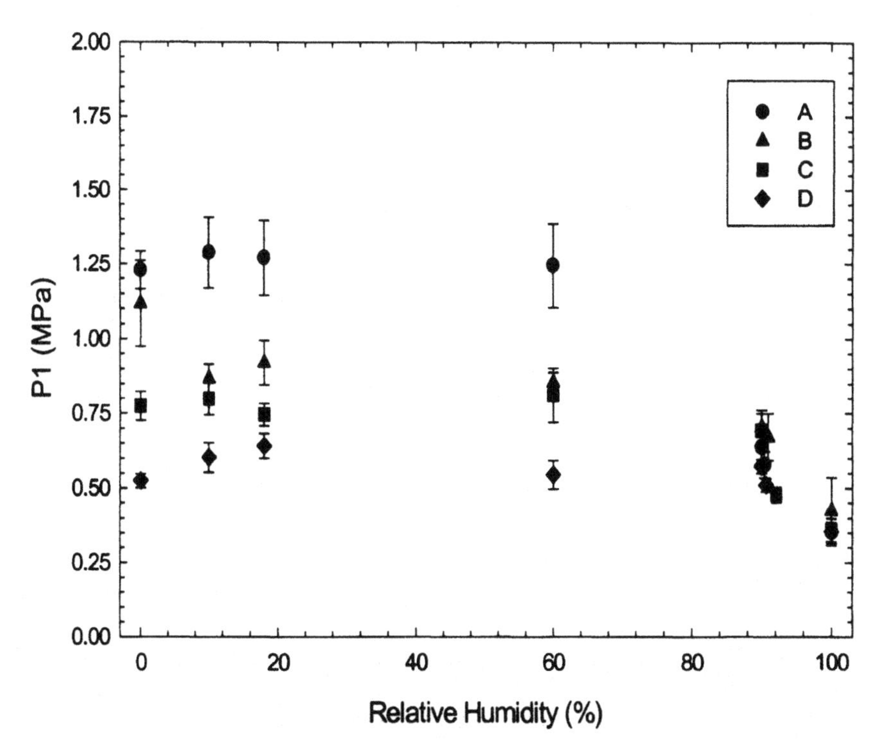

Figure 9. Plot of the pressure at the onset of granule deformation (P1) as a function of relative humidity.

Discussion

Moisture content has a significant effect on the deformation of the granulate in the binder systems tested. Moisture acts as a plasticizer for many organics, reducing the glass transition temperature, thus making the organic more deformable (or softer). When moisture is removed from the granulate, P1 is seen to increase, and when the granulate is rehydrated at high relative humidity, it is possible to reintroduce physical water into the granulate that again serves as a plasticizer.

The results from Granulates A and B indicate that when all water is removed (at 110°C) prior to rehydration, the granulate must be stored at greater than 90% relative humidity to reintroduce sufficient water to make the granulate deform like the as-received material. In the case of Granulate B, there is an effect at lower relative humidity as demonstrated by the drop in P1 at 10% relative humidity.

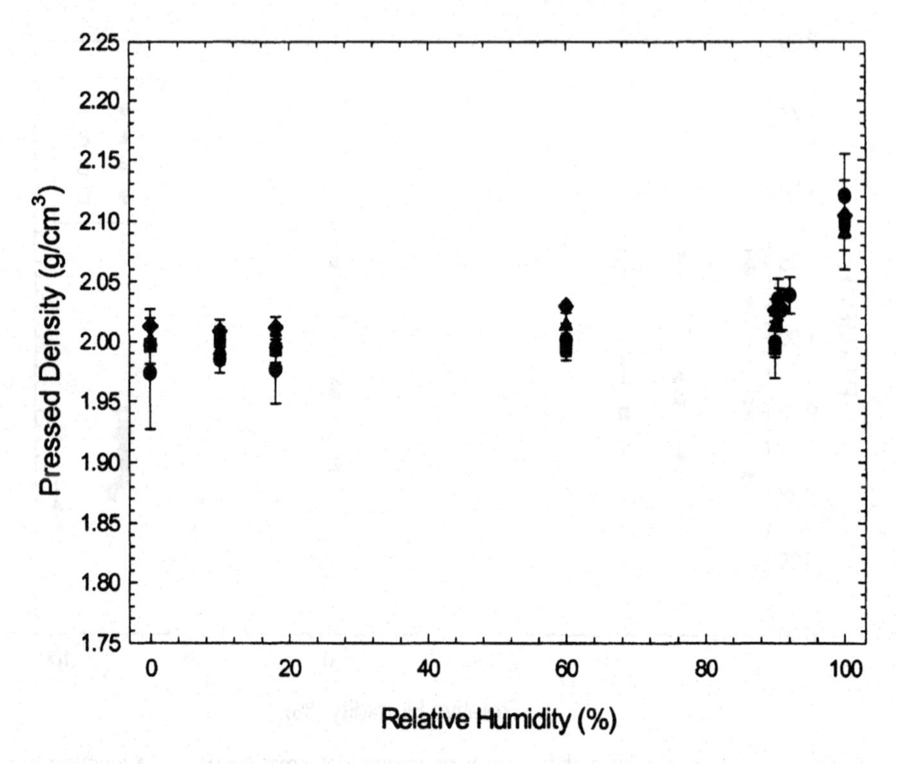

Figure 10. Plot of the pressed density as a function of relative humidity.

When the samples are stored at 100% relative humidity, the properties of the rehydrated granulate are seen to exceed those of the as-received material. The problem with storing granulate at 100% relative humidity is the growth of secondary organics (i.e., fungi) within the granulate.

Effect of Granule Size on Compaction: Moisture Content Distribution Argument

Experimental Procedure

Samples of granulate were sieved into size fractions ranging from 600 µm (30 mesh) to 45 µm (sub-270 mesh). Each size fraction was tested for moisture content by heating the sample to 110°C and measuring the weight loss. The remaining granulate from each size fraction was used to generate compaction diagrams to determine the pressure at the onset of granule break-

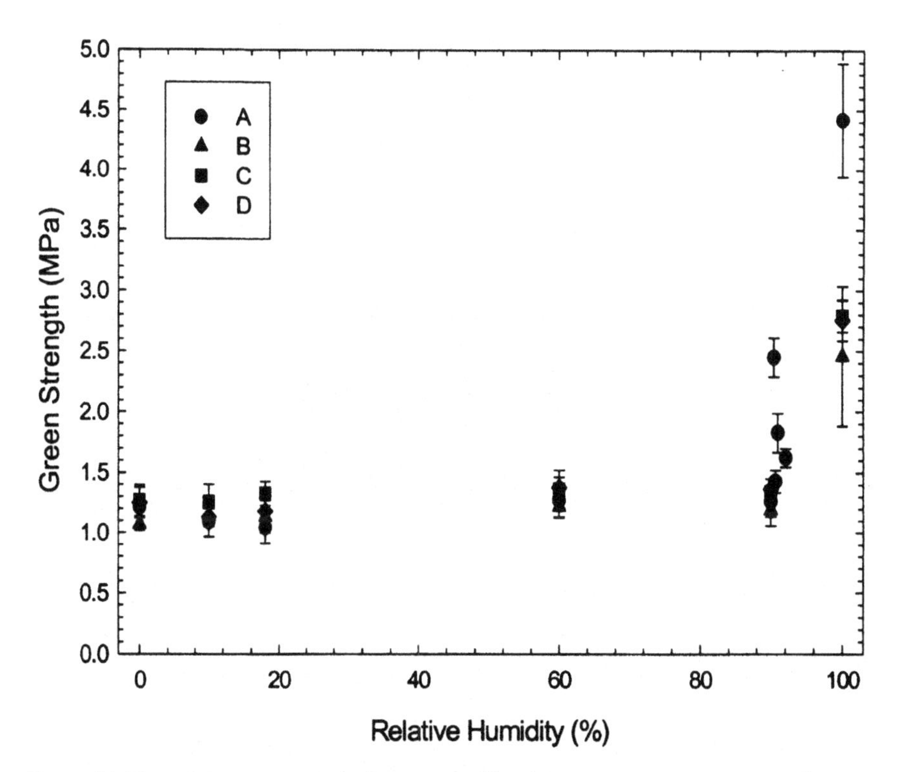

Figure 11. Plot of the green strength, measured by diametric compression, as a function of relative humidity.

down. The resulting pellets were tested for moisture content, pressed density, and green strength.

Results

The plots of the granule size distribution can be seen in Fig. 12 (Granulate A), Fig. 13 (Granulate B), Fig. 14 (Granulate C), and Fig. 15 (Granulate D). The average moisture content in each size fraction is also shown for each of the samples. There is a significant moisture content distribution in the granulate samples. The moisture content of each sample was tested at the time that the supersacks were prepared by Buffalo China. The average moisture contents, measured at the time that the granulate was prepared and during this study, are listed in Table III. There is a significant decrease in the average moisture content of the granulate after it was stored at the dry pressing facility, a facility that does not have any type of environmental

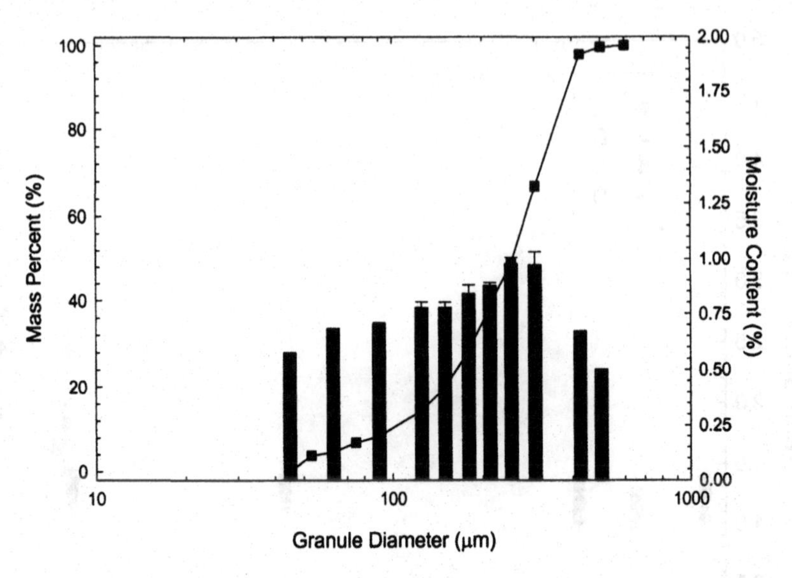

Figure 12. Granule size distribution and moisture content distribution for Granulate A. A moisture content distribution is seen in the granulate with high moisture content in the larger granules.

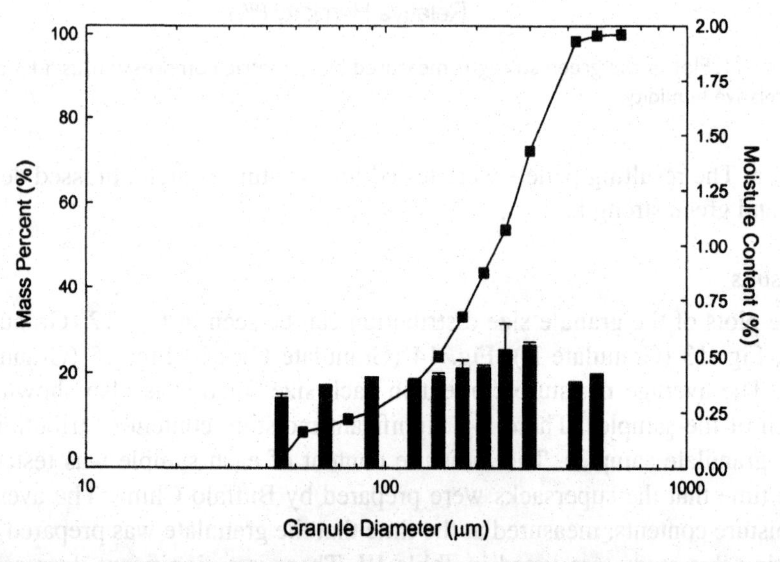

Figure 13. Granule size distribution and moisture content distribution for Granulate B. A moisture content distribution is seen in the sieved granulate, with the larger granules retaining more moisture.

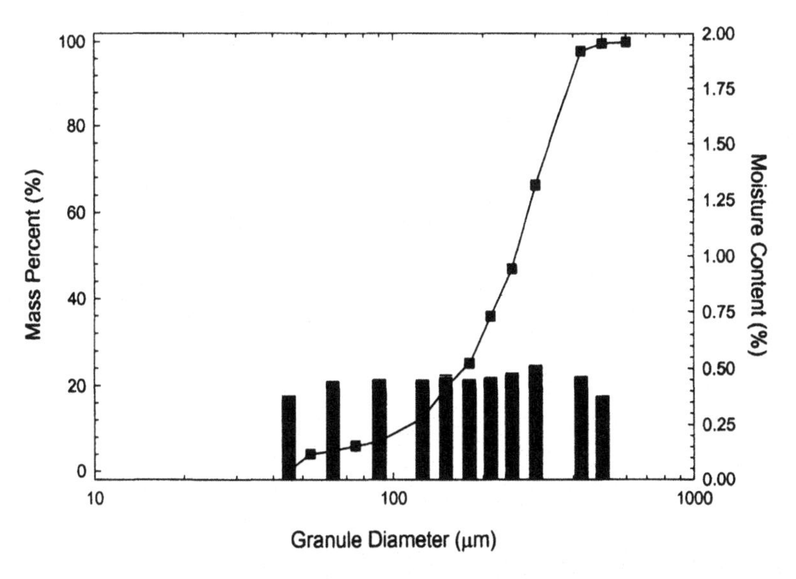

Figure 14. Granule size distribution and moisture content distribution for Granulate C. A small moisture content distribution is see in the granulate.

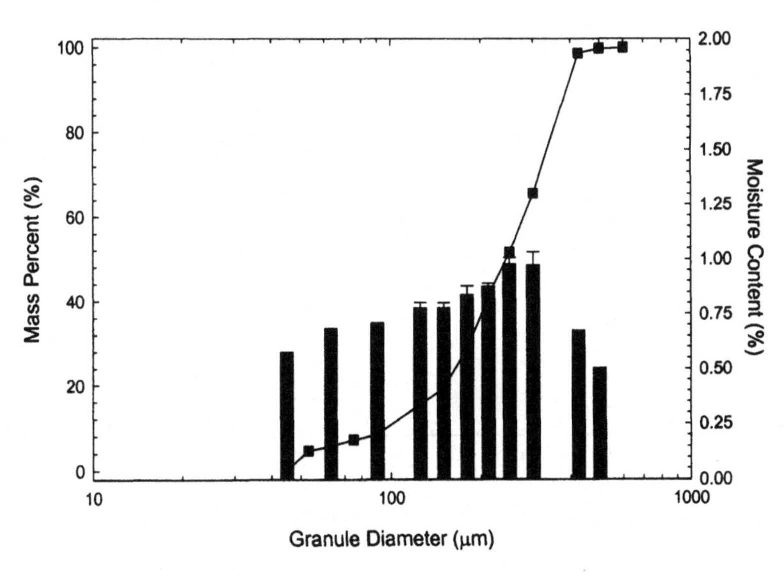

Figure 15. Granule size distribution and moisture content distribution for Granulate D. A moisture content distribution is seen in the sieved granulate, with the larger granules retaining more moisture.

Table III. Moisture content of the granulate in this study at the time that it was prepared by Buffalo China and at the time of analysis in this study

Binder system	Moisture content when spray dried* (wt%)	Moisture content at time of analysis† (wt%)
A	2.50 ± 0.56	1.04 ± 0.05
B	2.47 ± 0.68	0.66 ± 0.04
C	2.36 ± 0.27	0.66 ± 0.04
D	2.06 ± 0.51	0.79 ± 0.02

*Determined immediately after spray drying
†Granulate has been stored in sealed containers since June 2002

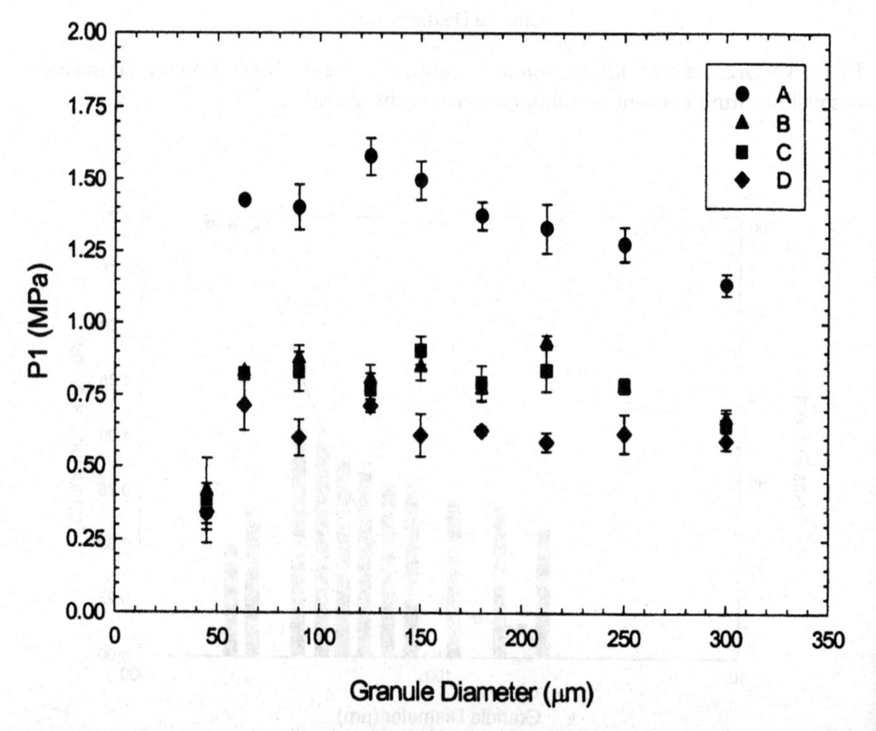

Figure 16. Comparison plot of the pressure at the onset of granule deformation (P1) for each binder system as a function of the granule diameter. The measured value for Granulate A is seen to be significantly higher than for the other binder systems in this study.

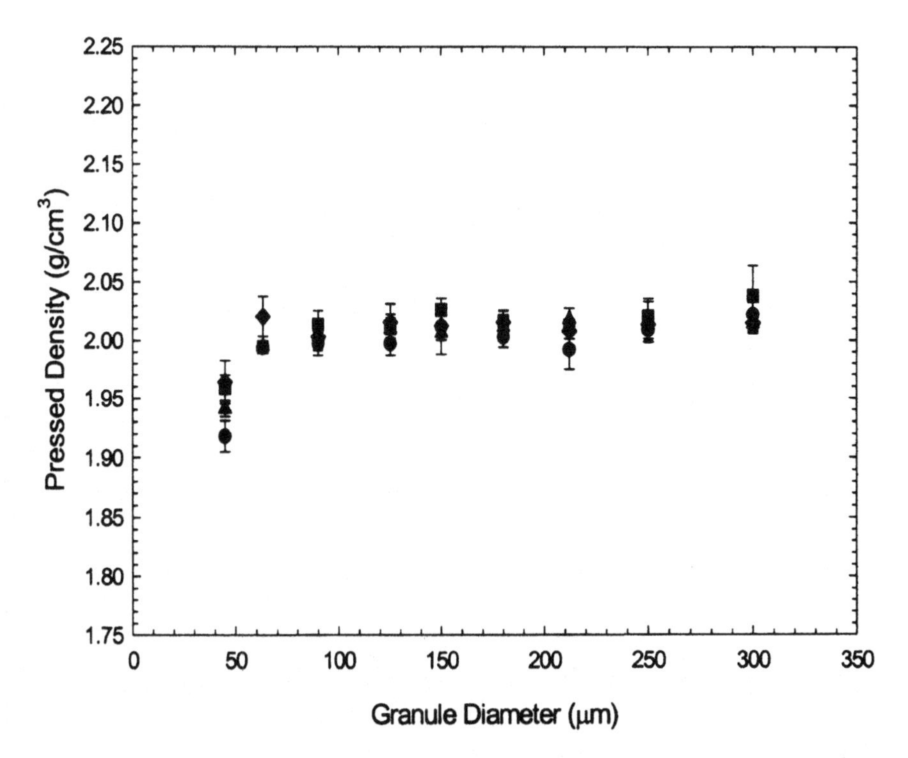

Figure 17. Comparison plot of the pressed density of pellets prepared for each binder system in this study as a function of granule diameter. Little difference is seen in the pressed density of the pellets prepared in this study.

controls. A sample of the stored Granulate A was returned to Buffalo China in January 2003 and subjected to their standard tests. The results indicate that there is no measurable moisture content in the sample (according to their standard practice).

The moisture content was also determined after pressing the pellets on the Instron (to determine the pressure at the onset of granule deformation, P1). There is not a significant moisture content distribution in the pressed pellets because of physical moisture from the atmosphere.

Comparison of the Granulate

A comparison of the results can be seen in Fig. 16 (P1), Fig. 17 (pressed density), and Fig. 18 (green strength) as a function of the granule diameter for each binder system in this study. The P1 value for Granulate A overall is much higher (neglecting the P1 value for the sub-230 mesh material) than

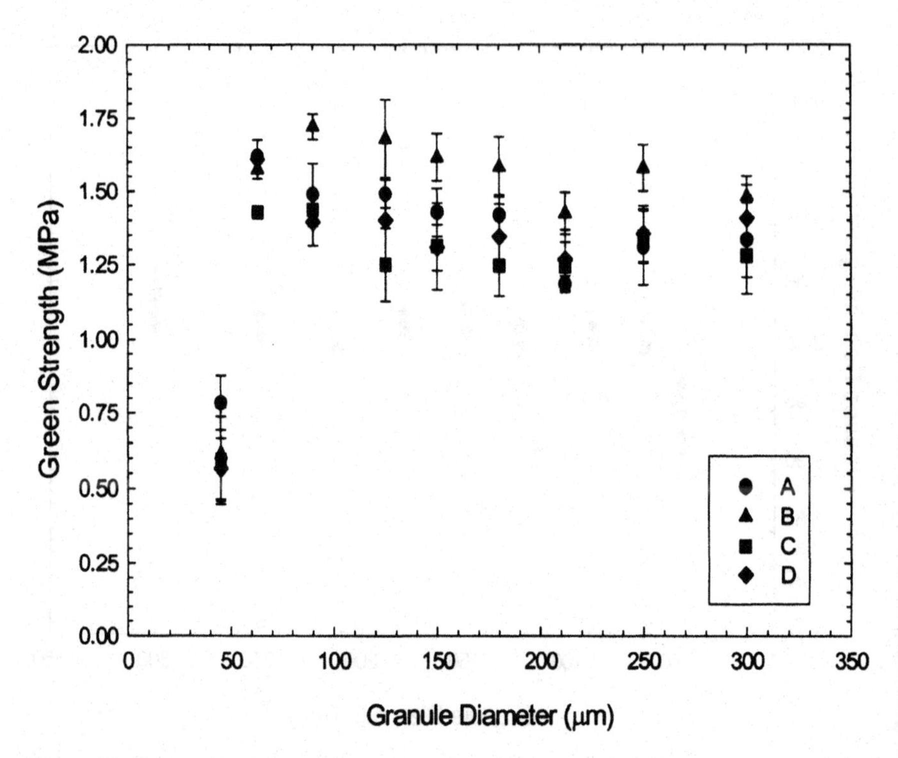

Figure 18. Comparison plot of the green strength, measured by diametric compression, for each binder system in this study as a function of granule diameter. Granulate B is had the highest green strength, followed by Granulate A. Granulate C had poor green strenth due to the soft nature of the binder.

for the other two binder systems. Little difference in P1 is seen between Granulates B and C. No difference in pressed density is seen between the binder systems in this study. The green strength of Granulate B is higher than for the other two binder systems. Granulate B had a lower P1 value than Granulate A, allowing the granules to deform under pressure (i.e., it is a softer binder system) and Granulate C is a very soft binder that is not effective at providing green strength to the compact.

Discussion

Only a small moisture content distribution is seen in the granulate samples tested. This is due to the storage conditions at the dry pressing facility. The granulate used in this portion of the study was stored for a period of 3 months under ambient conditions in unsealed supersacks. This resulted in constant fluctuation of the moisture content within the granulate in an effort to equilibrate with the surround environment. At the time that the granulate samples were taken, the relative humidity was high (greater than 80% in June 2002), and during the past 6 months of storage in sealed Nalgene containers, the samples have still not reached an equilibrium moisture content in all size fractions.

This is a consequence of the spray drying process. The finer granules dried earlier in the spray dryer and were subject to a thermal treatment that made the polymer hydrophobic in nature (i.e., hydroxyl groups were removed from the PVA structure, resulting in a carbon-to-carbon double bond). Larger granules retained moisture longer in the spray drying process and therefore remained cooler, allowing the polymer to remain more hydrophilic. The change in the nature of the polymer binder and the moisture content in each size fraction can be correlated to the change in the pressure at the onset of granule deformation.

While all size fractions (with the exception of the fines) were statistically pressed to the same density, there is a significant difference in the green strength of the samples as a function of granule diameter. Larger granules are seen to have a lower green strength due to the larger flaw size (the porosity remaining between deformed granules) in the compacts prepared with large-diameter granules.

Conclusions

Binder selection plays a strong role in the properties of spray-dried granulate. Higher P1 values were expected to result in higher measured surface roughness. Roughness was not seen to correlate with the pressure at the onset of granule deformation. Poor die fill has been demonstrated to increase the measured surface roughness by hindering good compaction during the pressing process. By improving the die fill prior to compaction using the vertical dry press, the surface roughness was reduced despite the higher P1 value of the granulate.

Moisture content has been verified as an important aspect in improving compaction. The presence of physical moisture in the granulate serves as a plasticizer that softens the granulate and reduces the P1 value. Higher moisture content (above 60% relative humidity) results in compacts with a higher green density and higher green strength when prepared in a hardened steel die.

Granule size has been demonstrated to have a small effect on P1 in the compaction diagram, neglecting the extreme fines (the sub-270 mesh fraction), which exhibited poor compaction for all binder systems. A moisture content distribution was observed in the granulate prepared with binder systems that thermally degraded during the spray drying process, despite storage in a sealed container for 6 months prior to evaluation. This moisture content distribution results in granulate with a range of P1 values for a single binder system, with the lower moisture content fractions exhibiting a higher P1 value.

Metal Marking of Dinnerware Glaze: Correlation with Friction and Surface Roughness

Hyojin Lee, William M. Carty, and Robert J. Castilone
Whiteware Research Center, New York State College of Ceramics at Alfred University, Alfred, New York

Although metal marking behavior of dinnerware glaze has been a persistent problem in the industry, a clear understanding of the underlying mechanisms remains elusive. Metal marks are defined as dark lines, often accompanied by damage in the glaze, caused by the deposition of metal during the use of metal utensils. It has been reported that certain types of glazes metal mark to a greater degree than other glazes. In this study, the mechanism of metal marking generation is proposed based on microstructural observations and friction measurements of matte, gloss, and zircon-opacified gloss glazes. Zircon crystallization was also evaluated to verify the effects of crystallization on the glaze properties including surface roughness and friction against metal. Severe metal marking was found in the zircon-opacified glaze; no glaze damage was observed in the matte and gloss glazes. Zircon increases the glaze's coefficient of friction, by the presence of protruding micron-sized zircon particles, resulting in crack formation on the glaze surface and allowing metal particles to be imbedded.

Introduction

The goal of this study was to evaluate metal marking behavior of matte, gloss, and zircon-opacified dinnerware glazes. Based on previous studies, it was found that metal marking behaviors depend on the glaze surface roughness and friction level to metal.[1] Matte glazes were found to metal mark easily at low applied pressures; however, the marks on the matte glazes, although dark in color, were easy to clean, making metal marking less of a problem. Clear-gloss glazes have high metal marking resistance, which was attributed to their smoothness. Zircon-opacified glazes tend to generate severe dark marks compared to the other glazes and have a limited ability to be cleaned after marking. These results indicate that zircon plays a major role in determining the metal marking tendencies in dinnerware glazes. In this study, metal marking on matte, gloss, and zircon-opacified glazes will be evaluated based on microstructural observation and the friction effects associated with zircon crystallization. To understand the opacified glaze surface in terms of friction to metal, zircon crystallization was also specifically studied.

Experimental Procedure

Zircon Crystallization Study

The composition of the raw base glaze is shown Table I. To represent the glaze compositions, unity molecular formula (UMF) notation will be used.[2] Suspensions containing zircon additions were wet ball milled for 2 h using alumina media, dried, and fired at 1230°C.

Table I. Composition of raw gloss base glaze

	Mol%	UMF
R_2O	4.0	0.2
RO	17.1	0.8
R_2O_3	10.3	0.5
RO_2	68.6	3.3

Quantitative X-ray diffraction (QXRD) analysis was performed to evaluate the zircon crystallization as a function of zircon added.[3] Figure 1 shows that zircon is primary crystalline phase; however, a small amount of quartz is also present. The mixed (312) reflection of zircon was used for the quantitative crystallization study to avoid preferred orientation effects. Zircon was added to the base glaze after firing to generate a calibration curve. The

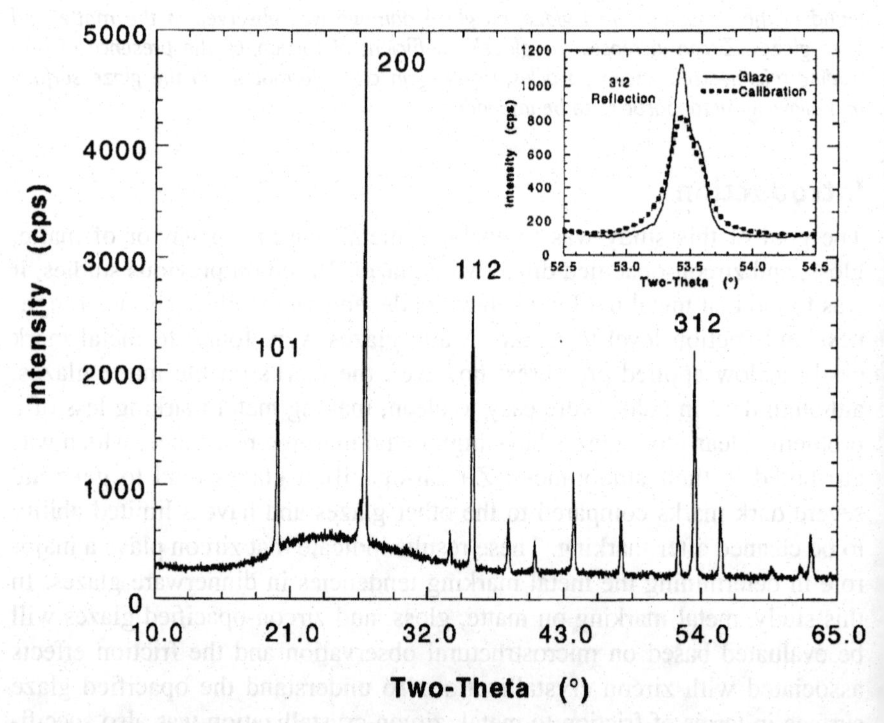

Figure 1. X-ray diffraction pattern for zircon in a glaze. The inset shows a closeup of the (312) reflection used for zircon quantitative analysis.

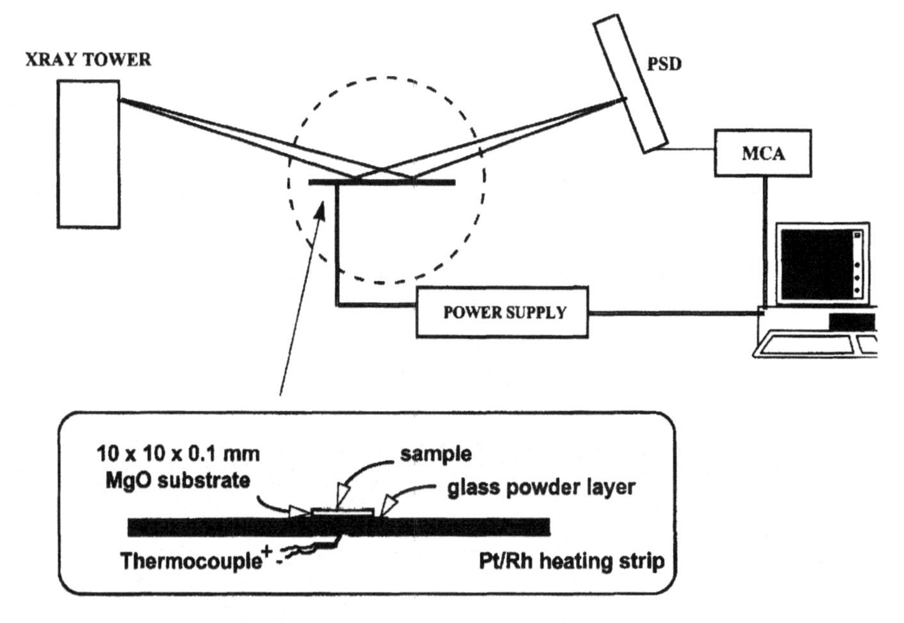

Figure 2. Schematic of the dynamic high-temperature X-ray equipment used in this study.

areas within the diffraction peaks were compared to the areas obtained in the calibration to determine the amount of crystalline zircon present in the glaze. Based on the closeup of the (312) reflection, both the $K_{\alpha 1}$ and $K_{\alpha 2}$ peaks are observed. Also noticeable is a difference in peak width between the calibration and glaze samples, which is attributed to a change in the particle size of the zircon particles during the firing process (i.e., the particle size increases with firing).

For the dynamic high-temperature X-ray diffraction (DXRD) study, two separate glaze batches were analyzed: a glaze containing 13 wt% zircon added to the standard base glaze, and a glaze consisting of zirconia and silica added to the base glaze. A schematic of the dynamic high-temperature diffractometer is presented in Fig. 2. The sample was mounted onto MgO single crystal attached to a platinum strip with glass. The temperature at the top of the MgO crystal was calibrated by melting substances with known melting temperatures on the crystal and observing the disappearance of diffraction peaks. Temperature calibration was accomplished using the melting points of sodium chloride and gold. A scanning position sensitive detector, capable of measuring a two-theta range of more than 10° at static position, was used to rapidly collect diffraction data.

Figure 3. The metal marking apparatus used to reproducibly place marks on dinnerware plates.

Metal Marking on the Dinnerware Glaze

Figure 3 shows the equipment to create the metal marking on the glazes. The machine is designed to produce the marks in a consistent manner. During operation, a plate is placed onto a turntable revolving at a selected speed. As the plate revolves, a knife is pressed against the plate with a constant pressure of 80 psi, corresponding to 7 lb (3.2 kg) of force distributed over the knife contact area, to produce a calculated pressure of 10 000 psi (69 MPa). These pressures are estimated to be similar to those observed in normal use situations.

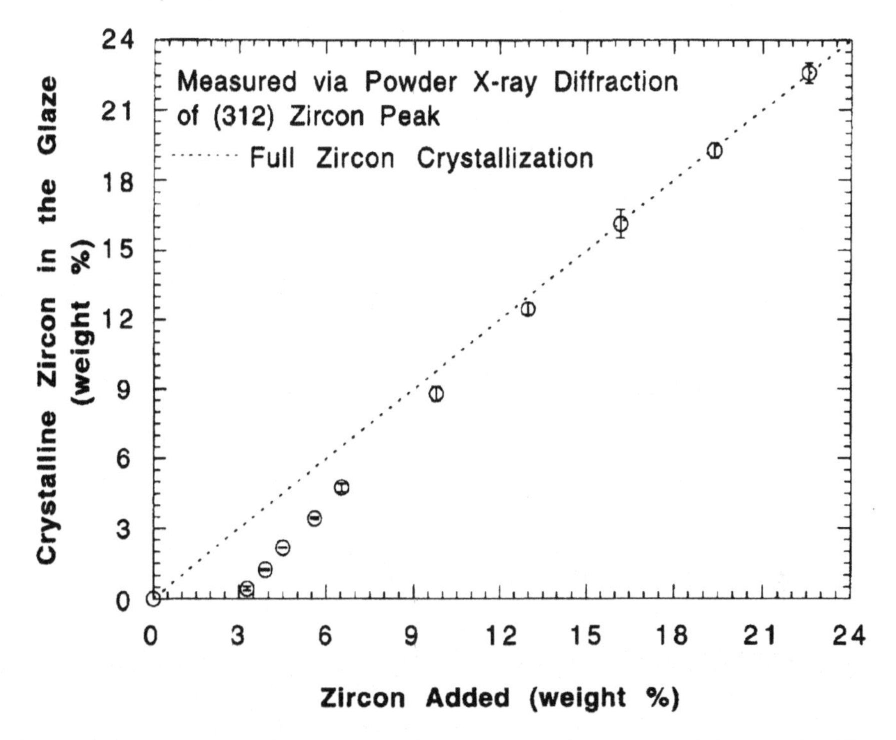

Figure 4. Amount of crystalline zircon in the glaze versus the amount of zircon added. Zircon is dissolved at low zircon additions but is fully crystallized at higher zircon contents.

Friction Measurements

The measurements were conducted using a friction apparatus consisting of a stainless-steel fork tine that was dragged across the glaze surface. The horizontal load required to drag the fork tine across the glaze was plotted against the applied vertical load; the slope of this line represents the coefficient of sliding friction.

Results and Discussion

Zircon Crystallization

Quantitative X-ray diffraction provided the amount of crystalline zircon in the glaze (see Fig. 4). For zircon additions between 0 and 3 wt%, most of the zircon is dissolved into the glaze. Between 3 and ~13 wt% zircon added, continuously more zircon is crystallized as the amount of zircon

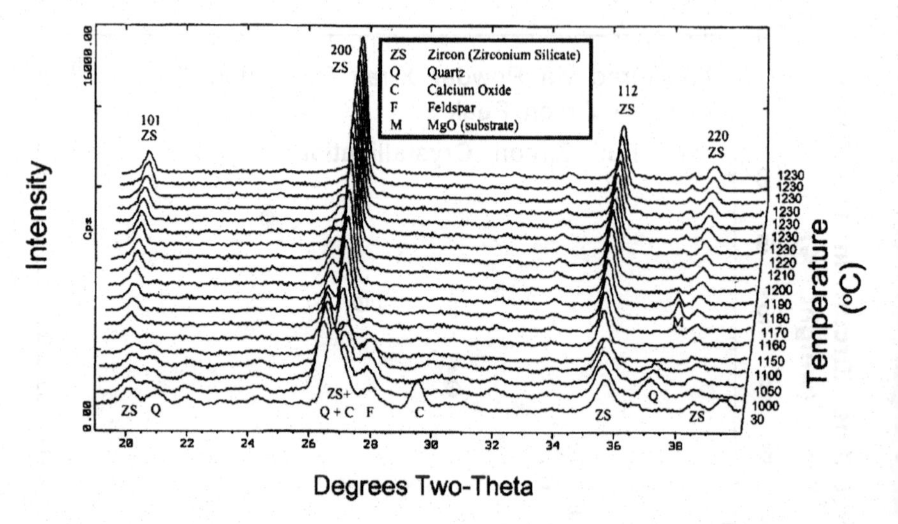

Figure 5. High-temperature X-ray diffraction pattern of the zircon-added glaze. Numbers above the zircon peaks represent the *hkl* reflections for zircon.

added increases. At zircon addition levels above 13 wt%, all of the zircon is crystallized.

The dynamic high-temperature X-ray diffraction analysis of the glaze with the 13 wt% zircon addition is shown in Fig. 5. The initial room-temperature run shows several raw materials present in the glaze batch, including zircon, quartz, calcium oxide, and feldspar. By 1000°C the calcium oxide is completely dissolved. The feldspar dissolves between 1100 and 1150°C, while the quartz remains until the maximum firing temperature and zircon remains as the crystalline phase in the glaze. However, during heat treatment, the zircon peak size and shape do not remain consistent.

The variations in peak area and width for the (112) zircon reflection with temperatures are summarized in Fig. 6. A sharp drop in the zircon peak area occurs between 1100 and 1150°C, indicating that zircon is dissolving. The decrease in crystalline zircon coincides with the temperature range in which the feldspar melts. The formation of a liquid phase resulting from the melting of the feldspar apparently triggers the zircon dissolution. As the temperature is increased, however, zircon begins to recrystallize in the glaze. With the recrystallization is a decrease in the peak width (full width half maximum, FWHM) indicating that the zircon particle size is increasing, or coarsening. This strongly suggests that the dissolution of zircon pro-

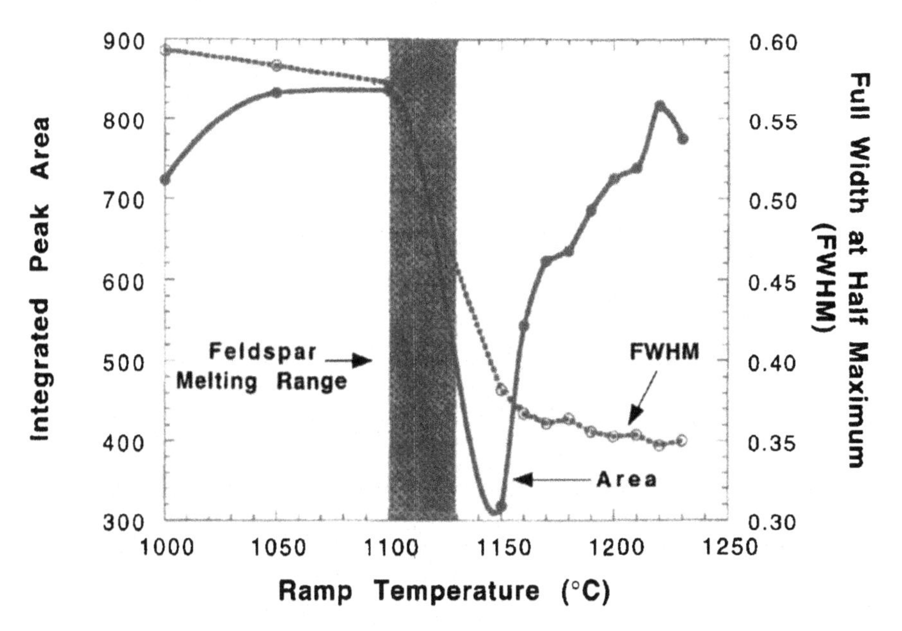

Figure 6. Peak area and peak width of the zircon (112) reflection versus temperature, obtained via dynamic high-temperature X-ray diffraction.

ceeds by the dissolution of small particles and the precipitation of dissolved zircon on the surface of larger particles.

A glaze of identical composition was prepared with zirconia and silica additions, but behaves differently compared to the glaze containing zircon. DXRD results (Fig. 7) indicate that zircon crystallizes during the heat treatment process, and post-firing analysis indicates that there is no zirconia left in the glaze after the firing process is complete. At 1180°C, the first zircon peak is observed in the glaze; also at this temperature, the low-temperature monoclinic zirconia phase diminishes, while the high-temperature tetragonal zirconia phase forms. Based on the quantitative analysis, the amount of zircon in the final zirconia-added glaze was only 9% crystalline, which was 4% less than the added amount.

Microstructures of two glazes were examined and compared using scanning electron microscopy (SEM). It is common to observe clustering of the zircon particles on the surface of the glaze (see Fig. 8). High-temperature microscopy of other glaze samples illustrated that during the heating treatment process, bubbles migrate to the glaze surface, displace zircon parti-

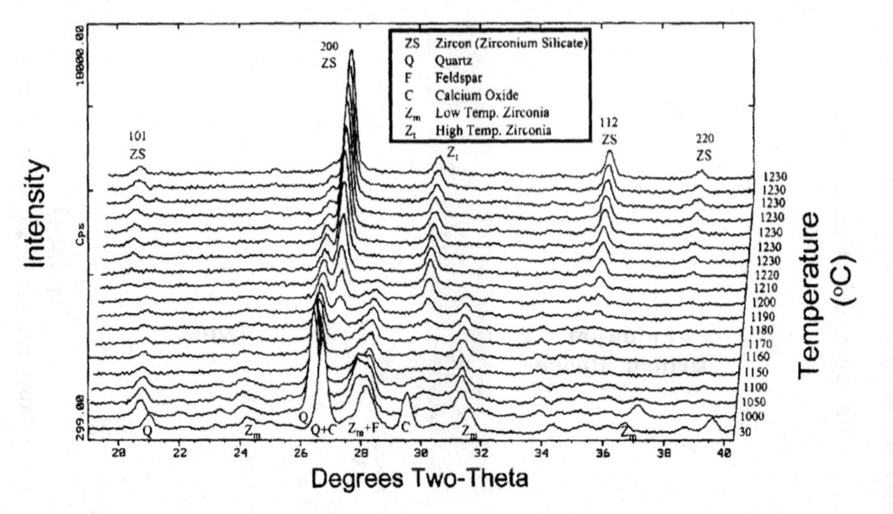

Figure 7. High-temperature X-ray diffraction pattern of the zirconia-added gloss glaze. Numbers above the zircon peaks represent the *hkl* reflections for zircon.

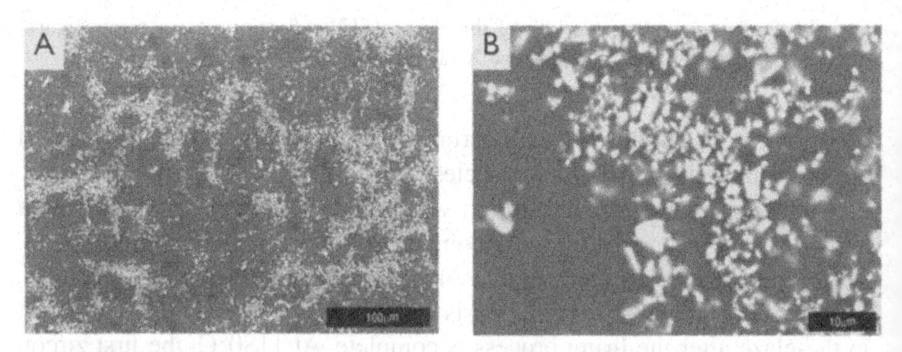

Figure 8. SEM photomicrographs of the glaze surface containing 10% zircon addition illustrating the zircon particle clusters and the zircon particle morphology: (a) low magnification (bar = 100 μm); (b) high magnification (bar = 10 μm).

cles, then burst, leaving a circular artifact of their passage. Figure 9 shows the SEM images of glaze prepared with the zirconia and silica additions, showing a needlelike morphology leading to a decrease in opacifier effectiveness.

Based on the zircon crystallization study, it was resulted that zircon remains dissolved in the glaze at low addition. Limited recrystallization occurs in glazes containing lower levels of zircon because of the complete dissolution of added zircon. High additions are required to obtain the opacification effects on the glaze.

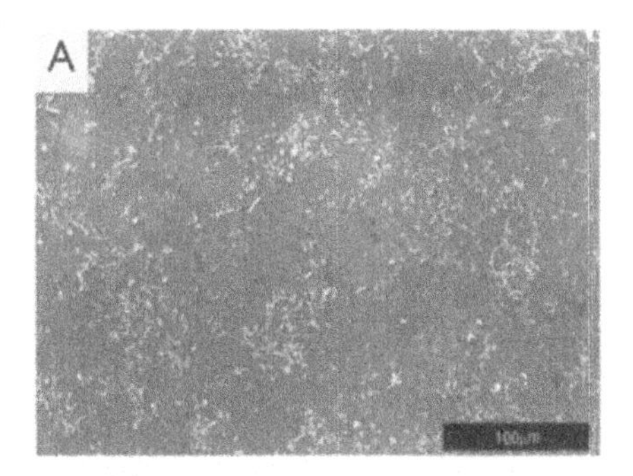

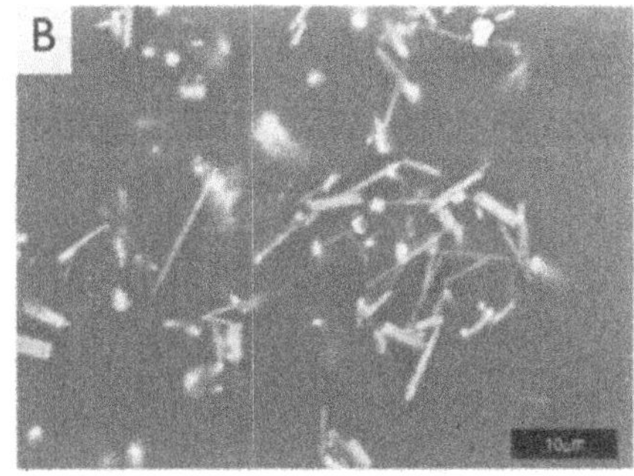

Figure 9. SEM photomicrographs of the glaze surface containing 10% zirconia addition, illustrating the distribution of zircon crystals and the high-aspect ratio zircon particle morphology: (a) low magnification (bar = 100 μm); b) high magnification (bar = 10 μm).

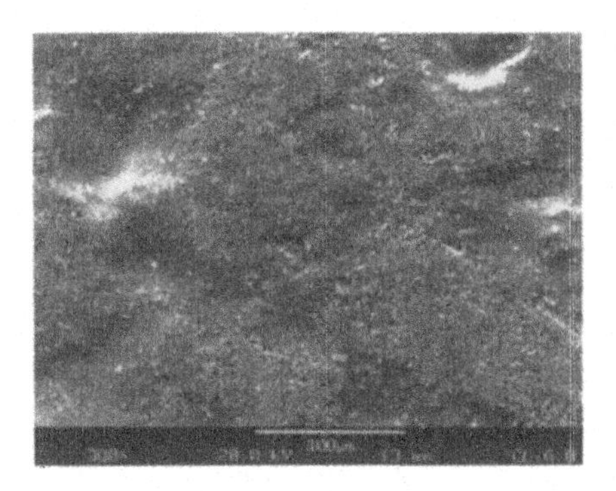
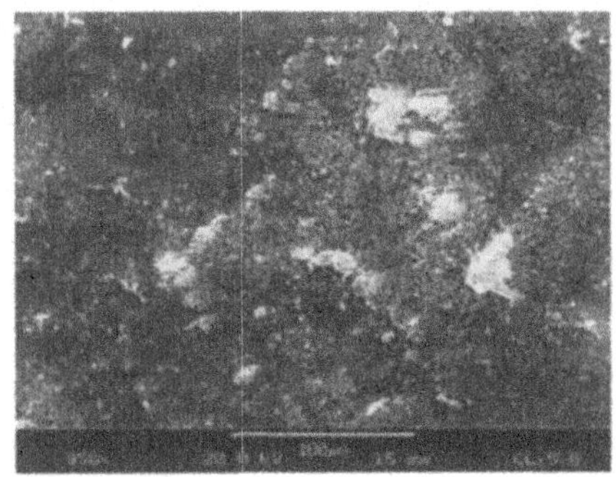

Figure 10. SEM photomicrographs of the matte glaze surface showing the unmarked glaze (left) and the marked glaze (right), demonstrating the large metal deposits on the high points in the metal-marked glaze (bar = 100 μm).

Metal Marking on the Dinnerware Glaze

The metal markings on the three glazes (matte, gloss, and zircon-opacified) were generated using the machine (Fig. 3) and examined using scanning electron microscopy (SEM). Figure 10 shows the SEM images of unmarked and marked matte glaze. The surface of a matte glaze is inherently rough, the result of partial devitrification of the glaze. Large amounts of metal were deposited in clumps on the glaze surface as the result of the abrasion of the knife by the textured glaze. The marks on the matte glazes appear as moderately dark but are discontinuous, presumably due to the marking tendencies mimicking the surface contours. Because of the necessary surface roughness for a matte glaze, it is impossible to eliminate metal marking on matte glazes.

Figure 11. SEM photomicrographs of the zircon-opacified gloss glaze surface showing the unmarked glaze (left) and the marked glaze (right), demonstrating the typical plow-shaped damage marks and metal deposits associated with the damaged areas (bar = 100 μm).

Figure 11 shows the SEM images of unmarked and marked gloss glaze. The glaze surface is relatively smooth, which prevents metal abrasion and therefore makes it difficult to mark. No glaze damage was observed, although a visible light mark was generated with multiple passes of the utensil. The metal marks were easily removed by cleaning. The gloss glaze demonstrates excellent metal marking resistance.

Zircon-opacified glazes with varied zircon additions were examined. According to previous work, the zircon addition must be at least 10 wt% to achieve adequate opacity. The glaze prepared with zircon additions less than 5 wt% behaved similarly to the base gloss glaze. However, crescent-shaped damage in the form of microcracks appeared in the glaze at zircon levels above 10 wt%. Large metal deposits are observed in the vicinity of the damaged areas, and in some cases appear to be embedded in the cracks. To the naked eye, the marks appear as continuous dark lines and are prominent.

Two types of metal marks were observed via microscopic analysis: marks with damage and marks without damage. The metal marks with damage occured only in the glazes containing high zircon levels. The coefficient of friction is an important factor in explaining the glaze surface damage. Higher values of friction will tend to reduce the normal load required to achieve crack formation. An increase in friction also tends to decrease the curvature of the cracks formed, an effect caused by a reduction in the symmetry of the stress distribution. Glazes having higher coefficients of friction contain a higher density of cracks as compared to lower-friction glazes. The coefficient of friction can be a dominant factor in the wear

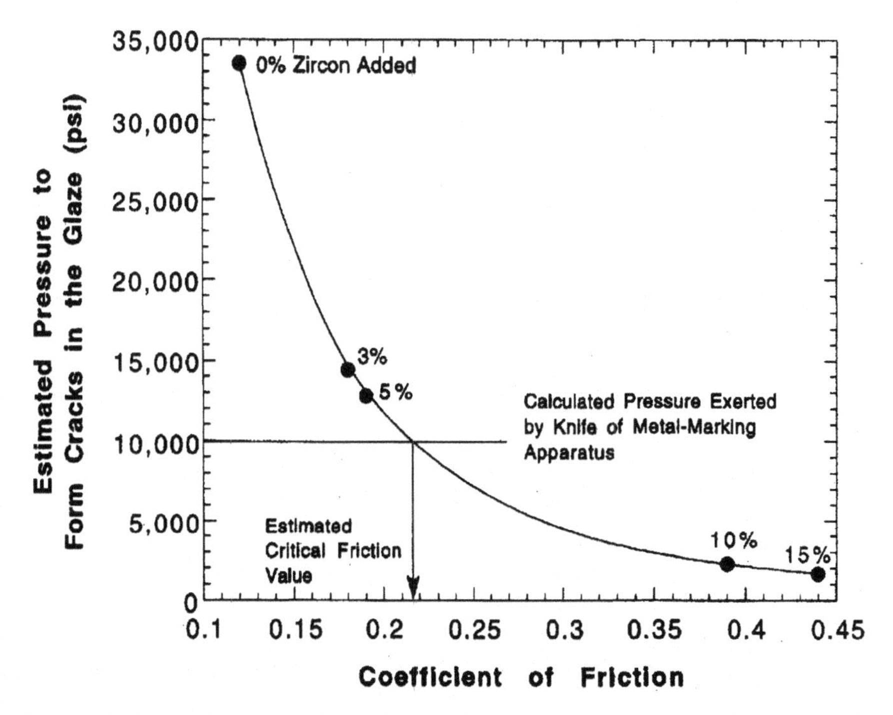

Figure 12. The reduction in the estimated force necessary to cause damage as the coefficient of friction increases. Zircon opacifiers increase the coefficient of friction, reducing the stress level necessary to cause damage and thereby increasing the ease at which severe metal marking can occur.

behavior of a glaze. The friction of zircon-opacified glaze against metal was examined using a friction apparatus. An increase in the coefficient of friction is present with increasing zircon contents. Figure 12 shows the calculated sliding loads necessary to produce cracking as a function of the coefficient of friction. A dramatic decrease in the pressure required to create cracking in the glaze occurs as the friction increased. Once the critical friction value is exceeded, cracks will form in the glaze at a specified applied pressure level. In the glaze studies, this critical friction value apparently was exceeded at the 10 wt% zircon level, and subsequently cracks developed.

Conclusion

Based on the zircon crystallization study, a high level of zircon is necessary to achieve an opacified glaze, but zircon additions increase the coefficient of friction, leading to surface damage of the glaze. When the surface is damaged, metal can become embedded in the glaze, leading to cleaning difficulties. Coarse marking, as observed with matte glazes, does not damage the glaze and is therefore easy to clean.

References

1. R. Castilone and W. Carty, "The Metal Marking Behavior of Stoneware Glazes," *Bull. Am. Ceram. Soc.,* **76** [3] 76–80 (1997).
2. W. Carty, M. Katz., and J. Gill, "Unity Molecular Formula Approach to Glaze Development," *Ceram. Eng. Sci. Proc.,* **21** [2] 95–109 (2000).
3. R. Castilone, D. Sriram, W. Carty, and R. Snyder, "The Crystallization of Zircon in Stoneware Glazes," *J. Am. Ceram. Soc.,* **82** [10] 2819–2824 (1999).

Development of Crystal Glazes

Jim Archer and Dave Schneider
Fusion Ceramics Inc., Carrollton, Ohio

Introduction

Crystalline glazes are characterized by the formation of crystals. They can be either microscopic or macroscopic, either in the glassy matrix or on the surface. The development of crystalline glazes allows us to achieve textural variations and not offer just the standard glossy or matte surfaces that are typically available. They contain materials that crystallize during the cooling period. The crystals usually occur in clusters, frequently covering considerable areas. Often, they are colored with ceramic oxides such as copper, cobalt, and nickel. The development of crystals depends on oxides, which saturate the glaze to form the corresponding silicate; the chemical and physical properties of the solvent glaze; firing and cooling conditions; and application.

Glaze Chemistry

Crystalline glazes are prepared by introducing oxides of the elements sodium, potassium, titanium, iron, and zinc to a glaze batch that has low alumina content. Other ceramic materials that can affect crystal formation are lithium oxide, calcium oxide, and molybdenum trioxide.

Certain glaze components are more conductive to crystal formation than others. To demonstrate this, we used three different glaze systems and fired them together to cone 5. The compositions used are shown in Table I. All three glazes are very soft and have a high flux content and low amounts of refractory oxides.

Table I. Glaze compositions used

	Glaze A	Glaze B	Glaze C
Zinc frit	80		
Sodium frit		80	
Lithium frit			80
Bentonite B	2	2	2
Silica	10	10	10
Zinc oxide	30	30	30

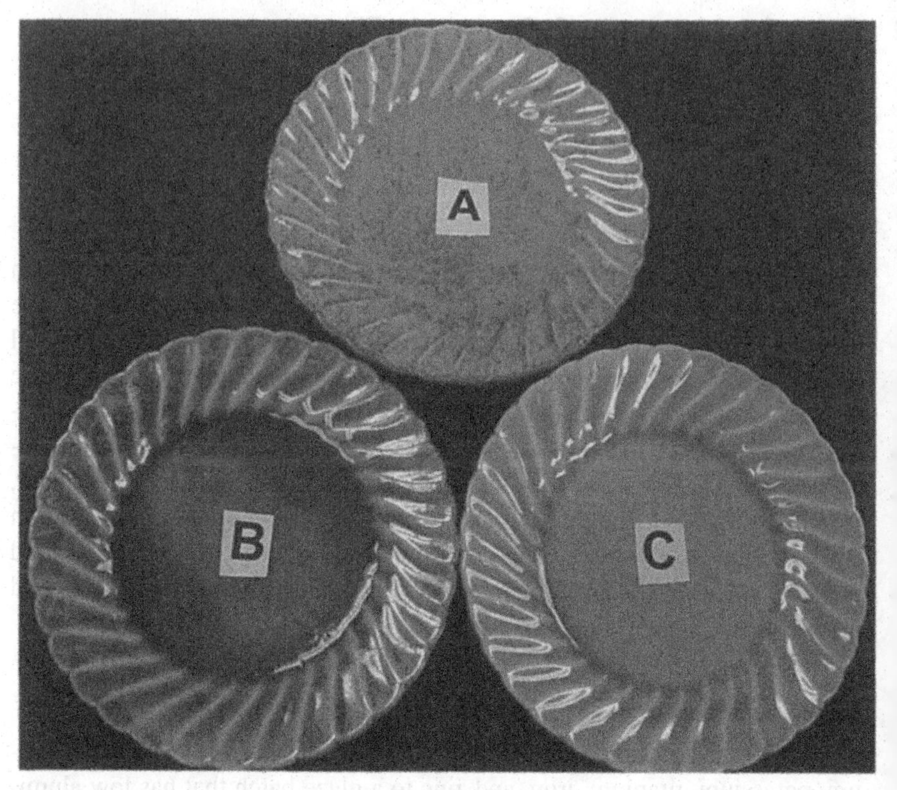

Figure 1. (a) High-zinc frit; (b) high-sodium frit; (c) high-lithium frit.

Figure 1 shows the formation of large fanlike crystals in Glaze B, which has the highest sodium content. These crystals have been identified as willemite ($2ZnO \cdot SiO_2$). We found no crystals in Glaze C that were based on lithium oxide. Glaze A developed fewer crystals than Glaze B.

Figure 2 is a closeup of the willemite crystals developed in Glaze B. NiO was used to enhance the crystal and give possible color variations of an otherwise transparent crystal. The blue color develops only in the zinc crystal and not in the surrounding glass matrix.

Figure 3 illustrates the image from the SEM of zinc crystal in Glaze B. The crystal is composed primarily of zinc, silica, and sodium. This explains why a glaze with high sodium content was more conductive to zinc silicate crystallization and produced larger and more pleasing crystals.

Figure 2. Zinc crystals in Glaze B.

Effect of Calcium on Titanite Crystal Growth

To illustrate how the addition of calcium can affect the amount of crystallization, we compounded two glazes: Glaze D (without overaddition of calcium) and Glaze E (with overaddition of calcium). Table II gives their compositions.

Because there is a small amount of calcium in the frit, Glaze D produced a few titanite crystals (Fig. 4). However, in Glaze E, with the over-addition of calcium, the crystals were so numerous that they gave the glaze a matte texture.

Figure 5 shows a closeup of the titanite crystals in Glaze E. They have been colored by cobalt oxide and have a needlelike appearance. Figure 6 shows their chemical composition as determined by a scanning electron microscope. These crystals are composed mainly of calcium, titanium, and silica.

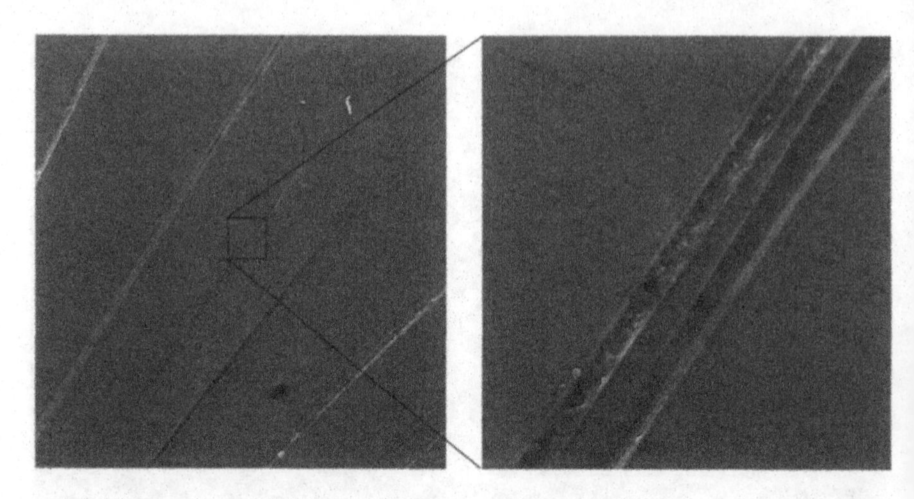

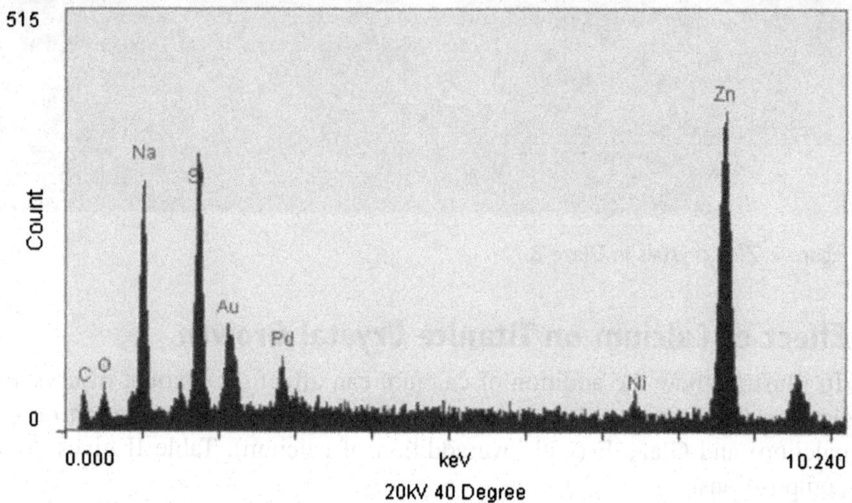

Figure 3. Images from the SEM of the zinc crystal in Glaze B.

Table II. Composition of Glazes D and E

	Glaze D	Glaze E
Alkali borosilicate*	80	80
Bentonite B	2	2
Silica	10	10
TiO_2	5	5
$CaCO_3$	0	10
CoO	0.2	0.2

*Frit contains 6% CaO

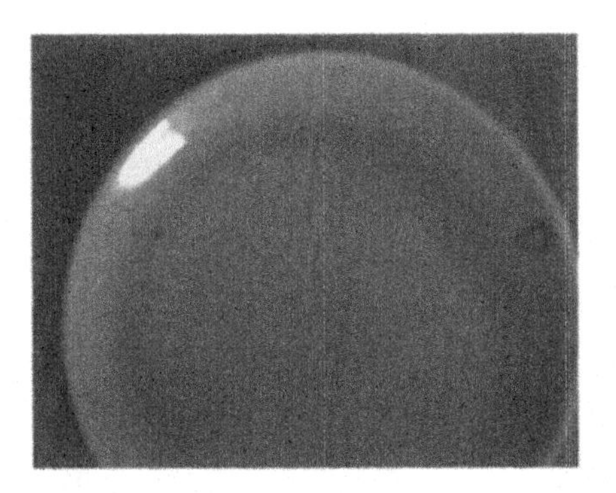

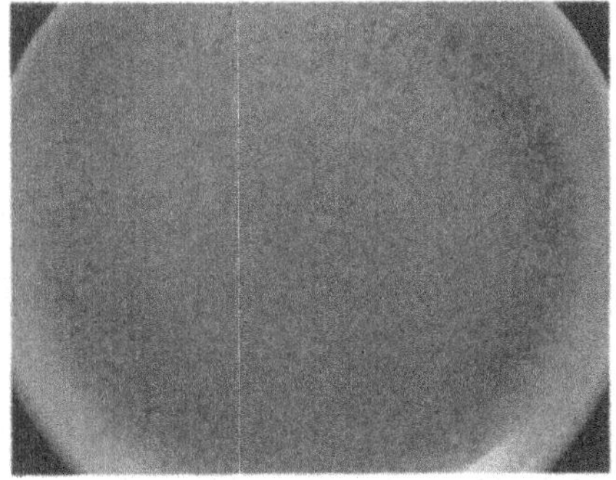

Figure 4. Glaze D (left) showed few or no crystals; Glaze E (right) showed numerous crystals.

Figure 5. Closeup of the titanite [OCaTi (SiO_4)] crystals in Glaze E.

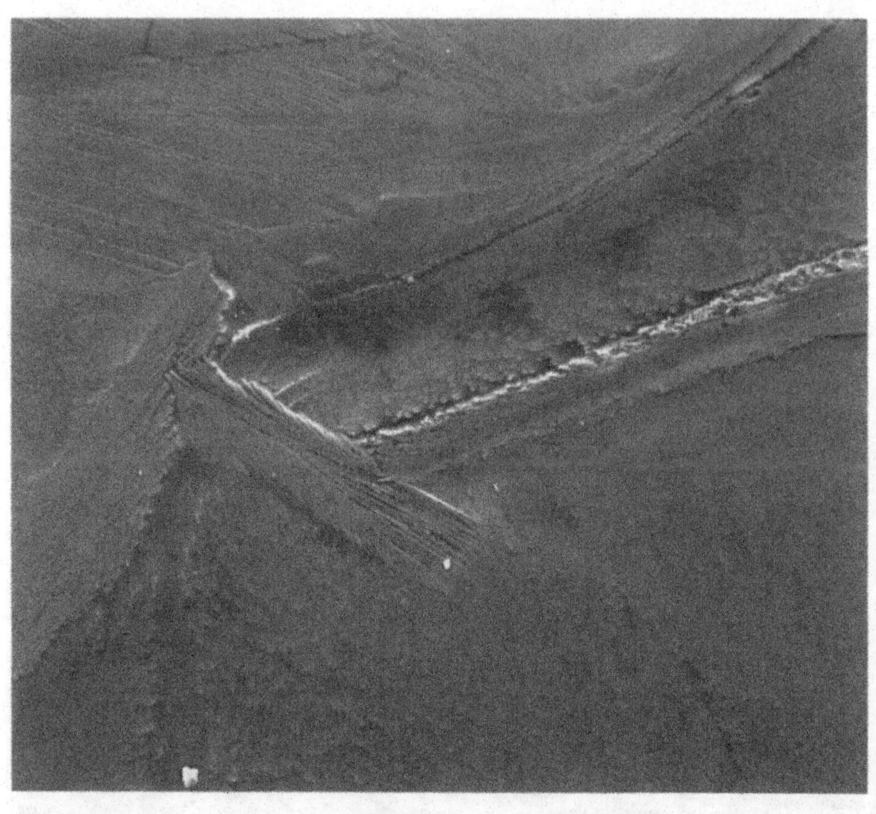

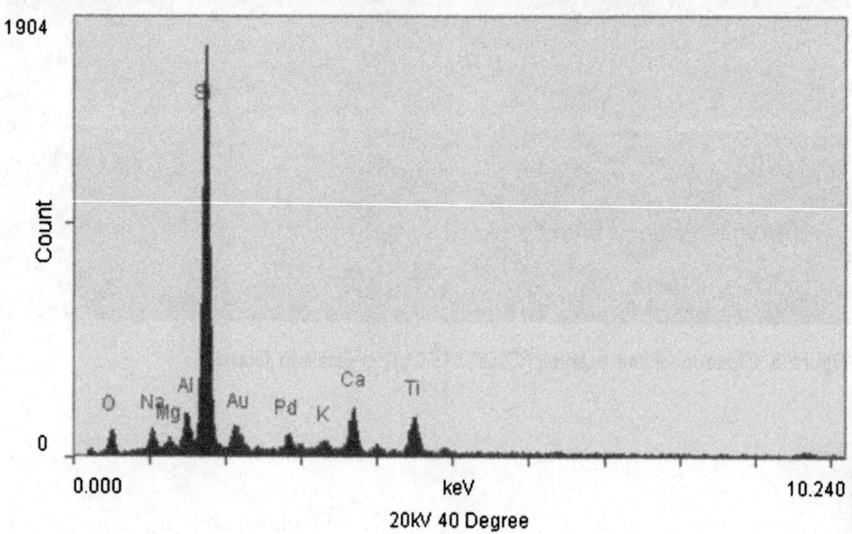

Figure 6. Images from the SEM of the titanite crystals.

Figure 7. (Left to right) +1% MoO_3, +3% MoO_3, and +5% MoO_3.

Effect of Molybdenum

MoO_3 is also known to enhance crystal formation. MoO_3 acts as a seed and creates nuclei that increase the number of crystals formed, as shown in Fig. 7.

Effect of Fired (Molten) Viscosity

Finally, the glaze must be favorable to the growth of the crystals, notably in viscosity. Because alumina increases the molten viscosity of a glaze, it can hinder or prevent the formation of crystals. We prepared three glazes using bentonite and kaolin as sources of alumina. Glaze F contains 1% bentonite, Glaze G contains 5% kaolin, and Glaze H contains 10% kaolin. Bentonite is a low-PCE (20) material and kaolin is relatively high-PCE (35) material.

Figure 8 shows that as the clay content increases, the amount of crystallization decreases. This is because the alumina in the clay is increasing the viscosity of the glaze and hinders the formation of crystals.

The composition of the glaze is important because it must serve as a solvent for the silicates, must contain the proper ratios and types of oxides for optimal crystallization, and must also be favorable to the growth of the crystals (i.e., have low viscosity).

Figure 8. Glazes with (top) 1% bentonite, (lower left) 5% kaolin, and (lower right) 10% kaolin.

Firing

Effect of Different Firing Rates

Crystals that form in this type of glaze are grown by controlling the decline in kiln temperatures and soaking the ware in the kiln. In this section, we will demonstrate the effects that different firing cycles have on crystal formation. First, we will show that the rate of rise in temperature does not affect crystallization. To do this, we compounded three different crystalline glazes and fired them in an electric kiln to 2185°F. They had different firing rates, but had the same soak times and soak temperatures. Table III shows the firing schedule we used. Figure 9 shows the results. The rate of rise shows no significant difference.

Table III.

Slower rate of rise	Faster rate of rise
2185°F at 270°F/h	2185°F at 1800°F/h
2185°F, 3-h soak	2185°F, 3-h soak
1956°F, 90-min soak	1956°F, 90-min soak
1949°F, 60-min soak	1949°F, 60-min soak
1897°F, 30-min soak	1897°F, 30-min soak

Figure 9. Rates of temperature rise vs. crystal formation. Top row: 270°F/h. Bottom row: 1800°F/h.

Effect of Soak (Rate of Decline)

Next, identical glazes were the fired to 2185°F with the same rate of rise of temperature but with different soaking times (Table IV).

From the illustrations shown in Fig. 10, we find that varying the soak times dramatically influenced the formation and rate of growth of the individual crystals. In fact, the glazes that had no soak formed no crystals. This is obviously a critical factor in developing crystal glazes.

To recap the results of our firing tests, we conclude that the rate of rise in

Table IV.

Firing cycle 1 (long soak)	Firing cycle 2 (slow soak)	Firing cycle 3 (normal cooling)
2185°F at 270°F/h	2185°F at 270°F/h	2185 at 270°F/h
2185°F, 3 h soak	2185°F, 1.5 h soak	Normal cooling
1956°F, 90 min soak	1956°F, 45 min soak	No ramp
1949°F, 60 min soak	1949°F, 30 min soak	
1897°F, 30 min soak	1897°F, 15 min soak	

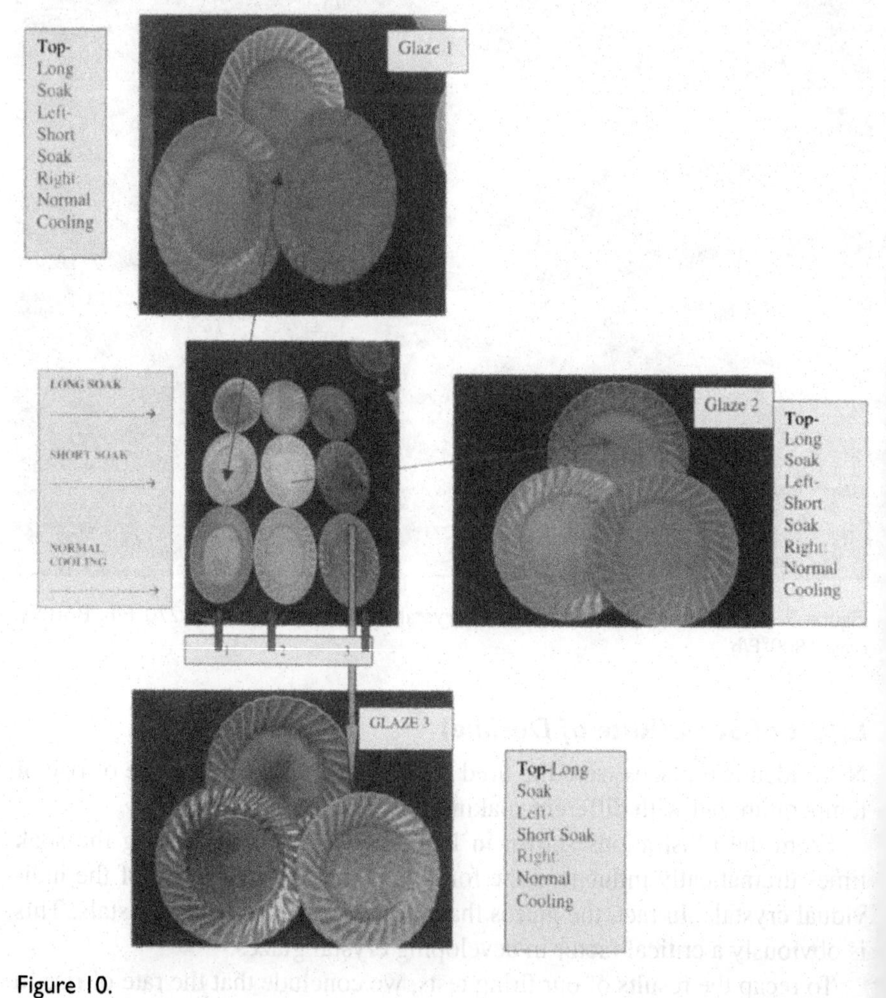

Figure 10.

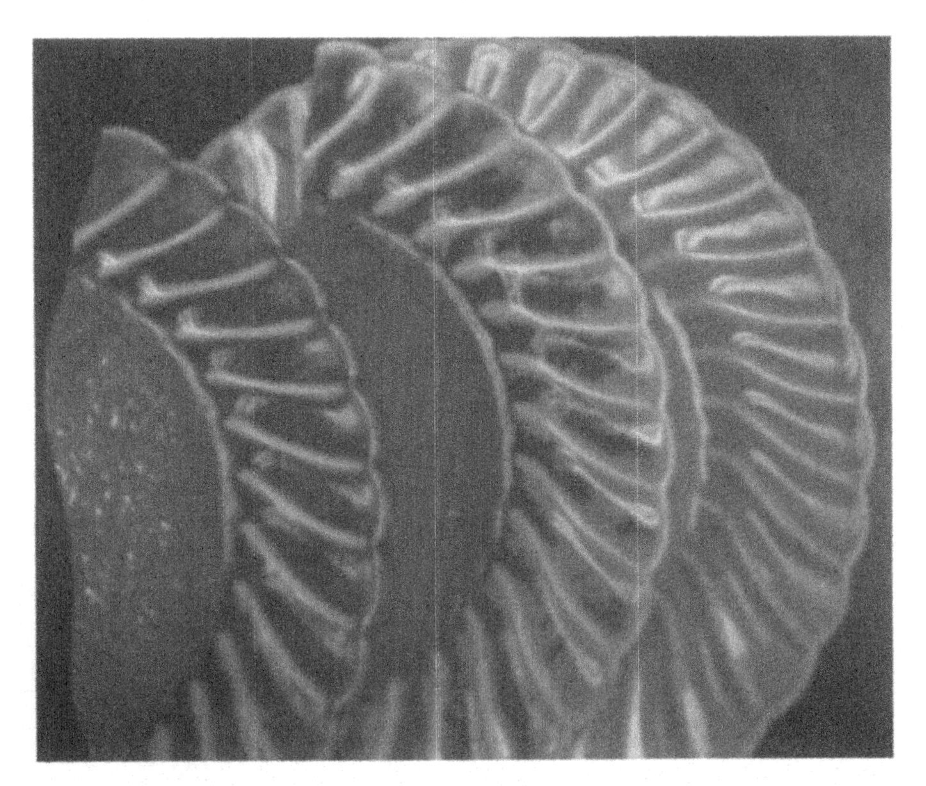

Figure 11. Left to right: 15, 10, and 5 g of application.

temperature does not require control different from that of ordinary glazes. However, the rate of cooling should be regulated with respect to the formation of the crystal nuclei and the rate of growth of the individual crystal formed. We also found that for each different crystalline glaze composition, there is an optimal time-temperature zone for maximum rate of crystal growth. The firing cycle can be complicated, so several tests should be carried out and the results noted, since each formula can vary and temperature maintenance may be different.

Effects of Application

It is useful to know how the thickness of the glaze layer can affect crystalline growth. To demonstrate this, we took a zinc-titanium crystal glaze and sprayed three different application weights (5, 10, and 15 g) on 6-in. plates. These were fired and evaluated for crystal formation. Figure 11 shows the results of our trial.

Figure 12.

Glaze thickness influences the fired viscosity and can increase or decrease the formation of crystals. In Fig. 11 it is clearly visible that more crystals are formed as the application weight increases. It is therefore necessary to determine the proper application weight of the glaze for optimal crystallization.

Influence of Coloring Oxides

Crystal glazes get their color through the addition of metal oxides such as cobalt, copper, chrome, vanadium, and other less common metals. Each oxide produces a characteristic color, which can be modified by other glaze components. As we mentioned earlier, nickel oxide in the presence of zinc produces blue crystals (Fig. 2).

The choice of oxides can also influence the rate of formation of the individual crystals. Some oxides, such as cobalt, iron, and manganese, reduce the maturing temperature of the glaze because of strong fluxing properties of these materials, while others, such as chrome and nickel, increase it.

Figure 12 shows a white zinc-titanium crystal glaze in the middle. To this glaze we added metal oxides to provide color and contrast.

Conclusion

The composition of the glaze is important because it must serve as a solvent for the silicates, must contain the proper ratios and types of oxides for optimal crystallization, and must also be favorable to the growth of the crystals. We also conclude that the rate of rise of temperature does not require control different from that of ordinary glazes. However, the rate of cooling should be regulated with respect to the formation of the crystal nuclei and the rate of growth of the individual crystal formed. We also found that for each different crystalline glaze composition, there is an optimal time-temperature zone for maximum rate of crystal growth. In addition, it is useful to know how the thickness of the glaze layer can affect crystalline growth. Coloring oxides can be useful in enhancing the color, texture, and design capabilities of crystal glazes.

Hotel China Glaze Reclamation

Michael Tkach
Homer Laughlin China Company

Homer Laughlin China produces vitrified china dinnerware for the commercial and residential markets. Three glaze application methods are used: flushing, waterfalling, and spraying. Flushing and waterfalling produce very little glaze waste, while spraying produces significant quantities of glaze waste to reclaim.

Why reclaim or recycle? We reclaim to reduce the environmental impact of glaze waste and to reduce costs. Other waste streams that are recycled are plaster and refractories.

Four sources generate the glaze waste stream: the spray booth, the foot sponger, the spindle washer, and the wastewater collection pit (Figs. 1–4). The spray booth is the largest generator of wastewater. During the operation of the booth, periodic washings are performed. At the end of the shift, the booth and dust collector are completely cleaned. The foot sponger generates the least amount of glaze waste but contributes to the overall quantity of water since it uses fresh water continuously. After the item is sprayed, the foot of the dinnerware is wiped on a wet sponge to remove the glaze. This allows the item to be fired without pins on a refractory setter. After each pass through the booth, the spindles are washed to remove glaze accumulated on the spindles. This water is taken from the collection pit and is recycled throughout the day. Finally, the collection pit recovers all of the water and glaze from the other sources and is emptied periodically during the shift and then after cleaning of the booth and dust collector.

The glaze waste water (1.04 g/cm^3) is collected in a 3400-gal tank (Fig. 5). This tank is equipped with high- and low-level detectors that turn the clarification process on and off automatically. The water leaves the holding tank, passes through an 80-mesh basket filter to remove trash (wood, rags, paper, grease, fired ware, refractories) and then to a five-fingered rare-earth magnet. The water is ready to be clarified.

The water enters the three-chamber clarifier at a maximum rate of 20 gal/min. In the first chamber, flocculant is injected at the bottom and mixed with a slow-speed mixer. The water flows to the top and over a weir to the second chamber and flows down and through a baffle into the third cham-

Figure 1. Spray booth.

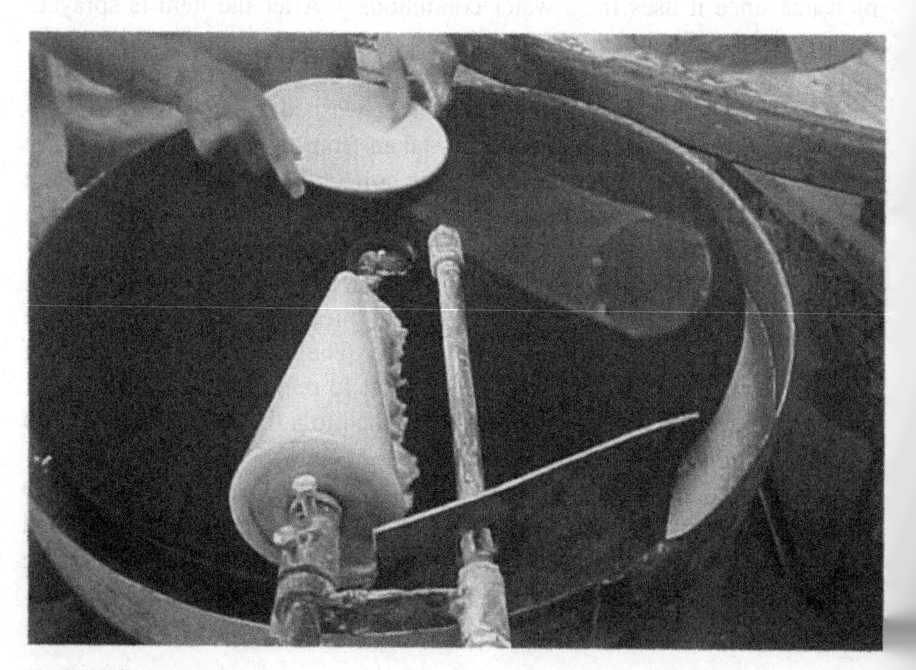

Figure 2. Foot sponger.

Figure 3. Collection pit.

Figure 4. Spindle washer.

Figure 5. Collection tank, filter, magnet, and clarifier.

ber. This serpentine flow allows for extended dwell time so that the flocculant can react.

The flocculated water leaves the third chamber and flows to a slotted weir that runs the full length of the clarifier and distributes the water equally. The flocs are large enough to begin settling. The water flows down to the bottom of the clarifier and then back up. On the way up, a series of closely spaced inclined plastic plates aid in settling out the finner flocs. The water exits the plate pack and leaves the clarifier. As the solids begin to build up, the water level in the clarifier rises. A float switch controls the level of the water in the clarifier and activates a pump that pumps the sludge (at 1.15 g/cm^3) to a holding tank (Fig. 6).

Agitators keep the sludge suspended. To dewater the sludge, a filter press is used (Fig. 7). An air-operated diaphragm pump is used to fill the filter press. The pump can complete the pressing in 1 h with a pressure of 80 psi. The average moisture of the cakes is 30.0%. Equivalent dry material from each press is 1782 lb.

Figure 6. Sludge tanks.

Figure 7. Filter press.

Figure 8. Blunger.

Figure 9. Screen.

Table I. Virgin vs. reclaim results

	Virgin		Reclaim
PSD <d30 (%)	100.0		97.7
SSA (m^2/g)	6.074		5.579
Color		no difference	
Gloss		no difference	
Thermal expansion		no difference	
Surface		no difference	

Table II. Economic impact of reclaiming glaze (2002)

Virgin glaze produced	2 048 000 lb
Reclaim glaze produced	565 198 lb
Glaze cost savings (565,198 lb @ $0.12/lb)	$67 823
Disposal cost savings (565,198 lb)	$10 368
Total savings	$78 191

The filter cakes are placed onto an inclined conveyor which takes the cakes to a blunging tank (Fig. 8). The tank is 45 in. in diameter and 41 in. high, and contains four baffles. Sixty-nine gallons of water are added to the tank along with the cake from one press. The cake is dispersed via a high-speed dispersion mixer. After mixing, the reclaim glaze is adjusted for specific gravity and viscosity. The last step is to screen the glaze over a 200-mesh screen (Fig. 9) and the pass it through an electromagnet.

Properties of the reclaim were compared to virgin glaze. The results are shown in Table I. The sole reason for reclaiming the wastewater was to reduce costs. The economic impact that reclaiming glaze had in 2002 is shown in Table II. Not having to produce 565,198 lb of glaze and not having to dispose of 565,198 lb of glaze saved $78,191 in 2002.